MW01099116

ENGINES OF PROSPERITY

Series on Technology Management

Forthcoming

Idea Generation and Use
by F. Hull *(Fordham Univ., USA)*

The Knowledge Enterprise
Implementation of Intelligent Business Strategies
by J. Friso den Hertog *(MERIT, Univ. of Maastricht, The Netherlands)* & Edward Huizenga *(Altuition in Den Bosch, The Netherlands)*

Japanese Cost Management
edited by Y. Monden *(Univ. of Tsukuba, Japan)*

Organisation for Japanese Technology
Japanese Style of Management in the New Generation
by K. Okubayashi *(Kobe Univ., Japan)*, H. Shoumura *(Tokyo Metropolitan College of Commerce, Japan)*, M. Morita *(Kansai Univ., Japan)* & N. Kambayashi *(Kobe Univ., Japan)*

SERIES ON TECHNOLOGY MANAGEMENT – VOL. 1

ENGINES OF PROSPERITY

TEMPLATES FOR THE INFORMATION AGE

GERARDO R UNGSON
University of Oregon, USA

JOHN D TRUDEL
The Trudel Group, USA

ICP

Imperial College Press

Published by

Imperial College Press
203 Electrical Engineering Building
Imperial College
London SW7 2BT

Distributed by

World Scientific Publishing Co. Pte. Ltd.
P O Box 128, Farrer Road, Singapore 912805
USA office: Suite 1B, 1060 Main Street, River Edge, NJ 07661
UK office: 57 Shelton Street, Covent Garden, London WC2H 9HE

British Library Cataloguing-in-Publication Data
A catalogue record for this book is available from the British Library.

ENGINES OF PROSPERITY
Templates for the Information Age
(Series on Technology Management – Volume 1)

ISBN 1-86094-092-7

This book is printed on acid-free paper.

Printed in Singapore by Uto-Print

Dedication

To my brothers and sisters: Victoria (Titus), Trinidad Jr. (Doody), Renato (Nato), Esther (Loida), Thelma (Emy), and Rafael (Binky)

G.R.U.

To my wife, soul-mate, and friend, Pat

J.D.T.

Preface

Most knowledgeable, thoughtful people now concede that there has never been any time in history when the human race has faced so much change. It is, paradoxically, both the best and worst of times. Opportunity abounds, but the economic, societal, and technological foundations of the Machine Age are crumbling. Confounded by chaos and heavily pressured for results, most Western managers have no better ideas how to compete than to endlessly copy each other, cut costs, and buy up rivals.

Not *every* firm is declining, of course. The situation is much like the national "quality crisis" in the early 1980s, a time when the exceptional U.S. firms were still the best in the world, but the average company was markedly inferior to its global competitors. (In both eras, good profits were often being delivered by increasingly uncompetitive firms.) The U.S. still excels at innovation, so many small U.S. firms do well — until they get large enough to show up on the radar screens and are savaged by global predators.

"A Nation at Risk" noted that if a foreign government had imposed our present educational system, we would regard it as an act of war. Fully 67% of the population say the American Dream is getting further out of reach. Globalization impacts national politics, sometimes in ugly ways. Foreign lobbying and political "donations" make the very issue which founded our nation — taxation without representation — once again timely.

Our society is having trouble focusing on, much less coming to grips with, these issues, for two basic reasons. The first is that in an age of information overload, reality can be made elusive. Alvin Toffler, the futurist, now says, "The sophistication of deception is increasing faster than the technology for verification. That means the end of truth."

Many politicians, executives, corporations, and governments are exceptionally effective at misdirection and covering up unpleasant truths. It is hard to fix problems when much of what we believe to be so, is not.

Exacerbating this fog of confusion is the fact that one of the strongest forces in today's society is denial. Much as in Rome during its decline, there is a tendency to use events to divert attention from troublesome systematic issues. In an age where most experts agree that the crucial international conflicts are economic, the British note that, "America is addicted to wars of distraction." The War on Poverty and The Cold War are now irrelevant, persistent involvement in squalid third world conflicts wastes valuable resources and accomplishes little or nothing, and the War on Drugs has been extremely expensive and largely counterproductive, if not disastrous.

Clearly, the world is going though a major transition. When this transformation is completed, it will look very different. This upheaval will change everything, but the focus and maximum stress point is economic. In the future, world power and national prosperity will increasingly depend on the ability to compete in high value added product-market areas. The winners will develop new societal models for business, economics, government, and education.

Most important is that the new basis of competition is knowledge. The strategic resource today is empowered minds and unique knowledge. Where nations once fought wars over trade routes and natural resources, future conflicts will be over the ability to produce and market uniquely valuable products. As never before — and in direct conflict with Machine Age models — human spirit, talent, education, and intellectual property protection will become central to prosperity. It is a time for renewed character, leadership, and spirituality, but these are scarce.

Those societies that can effectively apply knowledge for economic gain will prosper. Those that cling to Machine Age models and economics will fall to third world living standards. This trend is already well established, and a common strategy in smart nations is to seize the high ground, the "sunrise industries." (It is well accepted that the smartest people in Japan work in MITI. Few, however patriotic, accuse the U.S. government of being smart.)

A small number of key industries will constitute worldwide "drivers" of growth that allow high value added and strong barriers. Economic, technological, and social developments will change competitive patterns, innovations will upset current growth trajectories, and smart multinational and transnational firms can reinvent themselves. The dynamics that undergird growth, development and renewal are rooted in non-linear cost and demand patterns characterized by ever increasing returns and escalating growth. We call these the "Engines of Prosperity."

Competing in this new environment requires a different managerial mindset, not one that is fostered by mainstream theories. Unfortunately, such behavioral change is difficult, and very much so for those who thrived under previous models. Old Think beliefs and behaviors persist, and, in fact, still dominate. One common form of insanity is to keep repeating the same actions and hoping for different results. Detroit did this with large cars, Tektronix with analog oscilloscopes, Apollo with workstations, DEC with mini-computers, IBM with mainframes, and Washington DC and Eastern Europe with Megastate bureaucracy.

We do not offer "sound bite" or prescriptive solutions. Because our arguments may appear unfamiliar or complex at times, we present them in their entirety for perspective:

- Despite the imperatives to change, deep-seated conventional beliefs and assumptions are difficult to change. Managers use defensive behaviors — denial, inertia, scapegoating, and calibration to justify old thinking. Learning to succeed in the new environment calls for effective unlearning of old ways (Chapter 1).
- Business has entered a new age: a knowledge-based society. Machine Age templates are poorly suited to managing new knowledge. World power and national prosperity will now be determined by the ability to compete in global markets of the Information Age. This dooms the Megastate and traditional hierarchical silos. Leaders in winning

firms create new product-market spaces profitably under non-linear, chaotic situations (Chapter 2).

- One of the drivers of the Information Age is globalization. This refers to the deepening linkages and interconnectedness of economic activities on a worldwide scale. A new global mindset is needed, along with discarding old notions that foreign goods should be disparaged. Ethnocentric attitudes are ill-fated when applied to understanding, competing, or cooperating with new contenders (Chapter 3).
- The drivers of the new technology have shifted from government "mega-programs" to the commercial sector. Today's PCs exceed the power of Cold war supercomputers. Cryptographic technology is migrating to mass market PC applications. The Internet, developed to allow survivability during a nuclear attack, now offers hope for truly global commerce. (Chapter 4).
- Traditional and conventional approaches to strategic management, rooted in classical economics and strategy, are inadequate for the new age. Newtonian thinking, functionalism, and equilibrium are increasingly replaced by quantum thinking, holism, and ordered chaos. These provide the foundations for the *Engines of Prosperity* (Chapter 5).
- The new Information Age is impelled by industrial transformations brought about by the capability of digital technology. The roots of this industrial upheaval are grounded in five basic factors we call the engines: learning curves, demand amplification, technology generators, bandwagon and "lock-ins" and competitive imitation, innovation/disruption, and the unbundling of the firm's value chain (Chapter 6).
- In a world soon dominated by ordered chaos, the emerging management edict will be governed by paradox — when to build, when to destroy, when to harvest, and when to grow. As the yin-yang of managerial action, New Think templates force us to confront the open-ended nature of the future of any business (Chapter 7).
- Managers trained in action, reductionistic thinking, and narrow focus become very frustrated when trying to learn productive Information

Age behavior. The most dangerous time is when managers schooled in old think make gestures at change. Acts like ordered downsizing don't help. Reflection, practice and learning lead to deep conviction that makes change possible (Chapter 8).
- Today's contemporary business calls for a different type of leader — one who demands an unwavering commitment to change, learning, and unlearning. We should discard the "coyote-type" training in favor of that of the roadrunner. The strengths are different, and so are the mental processes and motivations (Chapter 9).
- Institutions matter. New institutions will develop to accommodate the requirements of an environment characterized by the *Engines of Prosperity*. Whether Americans use institutions to their advantage, or choose to ignore them, constitutes the final challenge (Chapter 10).

While our book is not oriented toward offering simple solutions, some broad guidelines might illustrate the spirit of our position:

- *Simplicity, while intellectually appealing, may not work.* Corporate America shows its penchant for the quick fix, sound bite solutions, and the next cute idea. Well, the world is complex, and, while others, and we, can cast it simplistically, there is no substitute for the hard work of understanding this new environment.
- *Complexity matters.* Complexity should not be confused with obtuseness. Chaos theory suggests that relatively "simple" patterns can be revealed in messy, chaotic settings. This is what we try to do with the "engines."
- *Think organically.* We have been socialized to analyze and segregate wholes. Yet, corporations are part of a larger world. Firms that have continuously outsourced are beginning to realize that something intangible gets lost in the process. We need to revert back to holistic thinking. Otherwise, one will not understand the "engines."
- *Reengineering management is not the answer.* It started with a simple idea — let's forget functional silos and concentrate on process, like selling. As a result, the reengineering industry was born, perhaps

creating more harm (about 40 million jobs have disappeared since 1993) than good. Reengineering does not work for non-linear or chaotic processes. It is a powerful medicine, but its limits and side-effects warrant more careful examination.

- *Recognize uniqueness*. There are severe limits to competitive imitation. Even Microsoft might not be able to do in 1998 what it had successfully done ten years ago. "Engines" illustrate why success in one company is not applicable to another.
- *Institutions matter.* Forget the criticisms levied at Asian cultures and work habits. Each country brings its own competencies to the competitive arena. We have our own in the United States. We should work harder at nourishing them. The "engines" force us to recognize our underlying strengths and weaknesses.

Engines of Prosperity is a joint product of an academician and a business practitioner, both of whom share a deep concern about the inadequacy of current models and practices. While we interviewed a number of managers from different knowledge-based industries, our book is not a product of "joint-research" in the traditional sense of hypothesis testing. Instead, we approached this book as a platform of shared ideas, borne out of our individual research and experience. Finally, while we draw heavily from chaos and complexity theories, our intention was not to write another book on this subject (there are a number of influential books that are referenced in our work). Rather, we have used such theories to explain the underpinnings of Information Age thinking that might help both business practitioners and academicians to better understand this new environment from the standpoint of their real-world experiences.

Gerardo R. Ungson
John D. Trudel

Portland, Oregon
1998

Acknowledgements

This book is a product of many studies, interviews, and consultations that spanned close to five years of intense collaboration. We met in a symposium on America's Industrial Policy, sponsored by Intel Corporation, in June 1992. While we spoke on different approaches to industrial policy, it was evident that we both shared some deep misgivings about what we, in this book, call Old Think. This led to meetings of auspicious beginnings, where John would attempt to frame patterns from his myriad business experiences, while Gerardo ("Buddy") would take John's numerous episodes and attempt to place them into theoretical categories.

In the process, we became indebted to many people who would read our work, present criticisms, or simply provide encouragement to go on. Buddy's interest in high technology began in 1982, when he was a visiting professor at the Haas School of Business, University of California-Berkeley. Corporate officers from Serafini Associates facilitated his field interviews with a number of high technology firms in the Silicon Valley. In Berkeley, he also met Professors John Zysman and Michael Borrus from the Berkeley Roundtable on the International Economy. From them, he developed a better appreciation of the institutional context of international high technology competition. He benefited from reading their papers and publications — their continuing influence is clearly evident in this book. We would also like to thank our students, Tak Yu Tong, Yssai Boussi, Ambrose Than and Henny Muliany, all from the Lundquist College of Business, for their assistance in our early research work.

Ideas for this book came about from people who devoted their valuable time for interviews. Companies included SSI, SEEQ, Intel, S-MOS, Hewlett Packard, LSI Logic, Signetics, Daisy Systems, Anthem

Electronics, Electric Scientific Instruments, Prometrix, Exel Microelectronics, Integrated office Systems, Fujitsu, Ashton Tate, Olivetti, and Philips Gloeilampenfabrieken. In extended trips to Korea, Samsung, Hyundai, LG Group, and Daewoo also provided valuable information on how their international strategies are shaped.

Of course, John has deep roots in knowledge-based industries, having started in Collins Radio Company, worked for several firms in the defense sector, and formed several new ventures. His last corporate tenure was a lengthy career with Tektronix, culminating in the position of Business Development Manager for their entire corporate Research and Development Laboratory, at the time one of the best in the world. His management consulting practice, under the aegis of The Trudel Group which he founded, brings him in daily contact with technologists and executives of first rate knowledge-based companies.

Trudel's clients over the past decade have included Bellcore, Cray, Exabyte, Intel, Lexmark, National Semiconductor, Sun, Illustra, Tektronix, and hundreds of others. All have added useful knowledge, even those (from large, established CAE-Link to tiny, stillborn Ty-Clamp) that did not survive. We offer special thanks to Will Swope and Pat Gelsinger of Intel, tough clients who provided high-bandwidth, mind-stretching interactions in the early years of The Trudel Group.

In addition Mr. Trudel serves as a National Examiner (one of five) for Product Development Management Association's "Outstanding Corporate Innovator" Award, which gives him the opportunity for detailed study of best practices across many industries. We are indebted both to the winners of this award — from John Fluke and Hewlett Packard and Kodak to 3Com and Herman Miller and others — and to the dozens of applicants who shared their insights under non-disclosure agreement. Finally, as a columnist for *Electronic Design* and *Upside*, and a crusader for preservation of patent rights, he has collected a network of thousands of fans, from all over the country and the world. All these experiences and contacts helped shape this book.

Early drafts were critiqued by Harry Lonsdale, who implored us to get to the point, while keeping the writing simple. Various portions of the draft were read by a number of Ph.D. students at the University of

Oregon, most notably Paul Fouts, who studied technology trajectories for his dissertation. Jim Goes and Seung Ho Park, now established professors in their own right, also provided numerous suggestions. Nic van Dijk from IME Consultants of the Netherlands and Alan Meyer from the Lundquist College of Business provided sharp critiques on the limitations of strategy in examining knowledge-based industries. Dennis Finnigan, consultant emeritus, helped us polish the book for CEOs and top management, while Tom Dagostino and Bill Sessa helped us make sure it addressed the issues of high tech middle managers.

We are especially indebted to Pat Trudel, who lovingly read every draft, making sure the book was upbeat for, relevant to, and readable by "normal people." Mary Mannin Morrissey helped provide spiritual inspiration and context, as well as the viewpoint of a mass-market author.

We are also indebted to the fine personnel from the Imperial College Press (ICP): Steven Patt and Yolande Koh who managed the editing and production, Aileen Parlane, who stepped in as Commissioning Editor, and to Colin McNeil who challenged his management to allow expanded promotion. We especially appreciate the creative freedom and support that ICP has provided.

Finally, we owe special thanks to our families in Eugene (Suki) and Portland (Pat), who cheered us on while we navigated through seemingly endless iterations of this book. John is grateful for Pat's support and encouragement during chaotic and challenging times: without her this book most likely would never have been completed. Any woman who would allow her honeymoon to be rescheduled for a business crisis assuredly deserves Sainthood. Buddy gives special thanks to his parents, brothers, and sisters in Santa Clara, for he would use every visit to the Silicon Valley as a pretext to come home and eat home-cooked, Filipino meals. They were wise enough to leave him the task of transcribing interviews while they attended to the arduous task of planning the next meal.

Gerardo "Buddy" Ungson
John D. Trudel

Contents

About the Authors

Gerardo R. Ungson is the Victor P. Morris Professor of Management at the Lundquist College of Business, University of Oregon where he teaches courses in business policy and strategy, organizational theory, and international corporate strategy. Professor Ungson's research interests cover strategic decision making, strategic implementation, and global strategies. His works have been published in numerous academic journals, and he has co-authored four books: *Decision Making: An Interdisciplinary Inquiry; Managing Effective Organizations; Chaebol: Korea's New Industrial Might;* and *Korean Enterprises: The Quest for Globalization.* He has taught extensively around the world, including Japan, Korea, and the Netherlands. He completed his undergraduate degree (Management Engineering) at Ateneo de Manila University, Philippines in 1969; an M.B.A. Management and a Ph.D. Business Administration from the Pennsylvania State University.

John D. Trudel is Founder and Managing Director of the Trudel Group (TTG), a high technology business development consulting firm that he established in 1988. TTG's clients range from new ventures to Fortune 100 companies such as Intel and Lexmark. Mr. Trudel enjoyed a successful early career as a technologist for Collins Radio Company, Sanders Associates, E-Systems and others. He worked for Tektronix where he played key roles in business venturing and new product development for several company divisions. Mr. Trudel writes a popular column, "Trudel to Form" for *Electronic Design* magazine, and for *Upside*. Mr. Trudel is the author of *High Tech with Low Risk.* He holds a BEE (cum laude) from the Georgia Institute of Technology and an MSEE from Kansas State University where he also did the course work for his Ph.D. He is a Certified Management Consultant.

PART I
ENVIRONMENTS AND MODELS

Chapter 1

Introduction

Key Themes at a Glance

"Old Think" — defined as the traditional, deep-seated, assumptions and beliefs on how to succeed in business — has become a poor guide to effective action in today's world. From a management perspective, these include the primacy of "top-down" thinking, undue emphasis on next quarter's bottom line, an unquestioned belief in sustainable competitive advantages, and "Band-Aid" crisis management practices.

From an organizational perspective, Old Think beliefs tend to emphasize functionalism over holistic thinking, the dominance of orderly processes, and a preference for programmed rote response over informed situational action. From an institutional perspective, this starts with what Peter Drucker has termed a "Megastate" approach, where government is the master, not the servant, of society.

Faced with external chaos, Western managers tighten control, preferring sanitary process and financial abstraction to the confusion and complexity of the real world. We like to say, "act, don't react" or "have plans and contingency plans" or "leave nothing to chance" or "never be surprised," and we wish that these mantras really helped.

We speak of "managing change," but we can't. Change is a wave to be ridden, not a machine to be managed. We speak of "managing time," but we can't. Time is a current or tide to be used and

harnessed, not managed. The fact is that in today's chaotic world we know so little and control so little. Control is the enemy of innovation and serendipity, the keys to prosperity in the new age. Which do you choose: predictability or prosperity?

We say that new viewpoints are needed to achieve prosperity. The dawning of global markets, the pervasiveness of technology, an enlightened work force, and new international competitors are why these must and will develop. We argue that such new templates, borne out of quantum physics, evolutionary economics, and chaos theory, demarcate the Information Age. In this Information Age context, we introduce the engines of prosperity as industry and organizationally based conditions that generate patterns of growth and renewal.

During the 1980s, computer-modeled business simulations and games, became fashionable at U.S. business schools. Teams competed against each other running simulated companies. Like their counterparts in the "real" world, students decided what products to produce, prices to charge, plants to build, and employees to hire. Anything legal was allowed, and the team with the best stock price won.

Early on, one of the authors played such a game in an executive training program co-sponsored by two prestigious East Coast Graduate Business Schools. This was a prototype executive program, so neither he nor the instructors knew quite what to expect. In anticipation of the end game, his team employed a "slash-and-burn" strategy: sell assets, cut investment, and fire the employees. At that time, the approach was revolutionary.

Their efforts were rewarded: they had the best profits and, hence, the best stock price. They were declared the "winner." The other teams were outraged. They protested, saying the strategy was absurd and unfair. The judges' ruling stuck, and the winners shrugged off complaints and accepted their prize. After all, it was just a silly game.

It is no longer a game.

Business corporations are engaged in a frenzy of downsizing activities — some help, but most don't, and many are ruinous. Since 1989, layoffs among U.S. firms have risen every year. Both the number and the extent of layoffs are on the rise. In 1996, AT&T announced plans (later softened) to slash half of its management in a single stroke.[1] These activities continue through the present, usually justified as effective cost containment to meet the requirements of stockholders. Few dispute the fact that many people and corporations have been hurt. A growing body of evidence shows that these methods usually don't help profits or productivity.[2] Even more disturbing is that recent research suggests that downsizing creates lasting damage to corporate creativity and morale.[3] The trend continues unabated and the popular comic strip Dilbert has become an icon for today's "clueless" management and cynical, fearful employees.

Old Think: The Dark Side of the Coming 21st Century

While some amount of downsizing is needed, the frenzy that accompanies "slash-and-burn" tactics, whether these tactics relate to a computer game or the real world, reflects a deeper malaise affecting U.S. corporations: the fallacy of Old Think. *Old think represents deep-seated but outmoded assumptions and beliefs about management methods to deal with external and internal changes confronting organizations.* In this book, our core argument is that Old Think managers instinctively and unconsciously frame problems and opportunities in terms we describe as Machine Age beliefs and assumptions. We also propose an alternative to Old Think that is based on quantum physics, evolutionary economics, and chaos theory — one that is more consistent with the requirements for operating and managing in a knowledge-based society.

Downsizing is simply one of many examples of this Old Think mindset. Interestingly, downsizing is often justified in the name of reengineering — so recently hailed as an all-powerful magic wand for corporate reform and renewal. The magic often failed, and most of the other fashionable management fads have yielded equally disappointing results.[4] After a decade of what *Wall Street Journal* has termed a "conspiracy of silence," there is now growing criticism of the poor results from the downsizing and merger frenzy.[5] Despite what is said about the need to reform, however, change is difficult primarily because Old Think is deeply entrenched in mainstream management theories and practice. Oftentimes, what is commonly held to be good management thinking masks incipient underlying beliefs that are outdated. Consider the following:

"What really matters is the bottom-line — the fallacy of short-term thinking."

Heightened competitive pressure is cited by managers as the key reason for downsizing activities. To attain expected financial performance, or the "bottom line," it is widely believed that costs have to be slashed. Labor costs, training, and new product investments are seen to be marginal costs, driving layoffs to improve the bottom line. What is less discussed is that the firms doing best in the U.S. invest in a loyal cadre of motivated knowledge workers, avoid layoffs, and invest in the future.

A related explanation is the pressure arising from quarterly evaluations, perpetuated in large part by Wall Street. Because corporations are punished for lower-than-expected rates of return, "slash-and-burn" tactics have become the norm.[6] The emphasis on the bottom line fosters a siege mentality and short term thinking. This was astutely voiced by one CEO who said, "If I waited any longer for the results of any investment, I may be gone from the company."[7] It's worse for division general managers.

Increasingly, the selection criterion for management is how many workers a manager has fired, and that rises in stock price accompanies massive layoffs — ominous signs. Upon his selection as CEO of IBM, Louis Gerstner Jr. was known more for having fired some 3,000 workers at R.J. Reynolds Tobacco Company than for prowess in the computer industry. He did as expected, and the stock price increased. A similar rise in AT&T's stock price accompanied the notice that 10,000 jobs were slashed. Were these firms really stronger or better after the layoffs? It's doubtful.

Box 1-1

> **"Get rid of abusive managers, even if they're stars. You can't have it both ways. Either an organization adheres to the team values it has chosen for itself or it rewards performance the old way."**
>
> Mort Myerson, Chairman and CEO, Perot Systems

It is unusual, but not unprecedented, to select leaders based on how many loyal troops they liquidated and how much of the realm they burned to achieve results. Russian historians would understand the approach fully. Still, even beyond the wreckage of downsizing is the societal harm and global trade disadvantage from the hollowing of entire industries. From consumer electronics, to aircraft, to computer memories and displays, Western industry has repeatedly abandoned lucrative growth markets and sacrificed good jobs to preserve short-term profits.

Our investment in research and development, especially in relation to international competitors, is in steep decline. More often than not, declining margins are cited as the reason. Bell Labs, inventor of the transistor and paragon of innovation, no longer does research, only incremental improvement.[8] Each year since 1994, the U.S. has come ever closer of selling out its patent system — another national treasure — to foreign interests.[9] The spate of "Barbarians at the Gates" type buyouts, mergers, and hostile takeovers have made the hollowing or dismantling of successful businesses and whole industries a common occurrence.[10]

"Managers plan and think, while their subordinates do the work — a misplaced solution out of context."

While its antecedents date from early history, the separation of "thinking" from "doing" was formalized by Frederick Taylor who, in 1917, sought to identify the scientific principles of work. In arguing for specialization, Taylor thought that managers perform their best when left alone to plan and chart the organization's future. Devoid of planning details, workers perform best when they are trained for highly specific tasks.[11]

While many persons have qualified this context as a relationship involving a supervisor and subordinate in a physical task (pig-iron steel technology), this "top-down" principle was interpreted to apply to all businesses at all levels of management. From this perspective, the essence of managing, as reinforced by mainstream theory, is to extract ideas from the heads of people at the top of the organization, and place them into the hands of those at the bottom. As span-of-control became a problem, process was added.

This mentality is reflected in populist theories about power, or how people move up the corporate ladder. As one progresses upward, one gains access to privileged information. But such information is rarely shared; in fact, having this information, or access to it, forms the source of one's power. Having this privilege creates dependence.[12]

Traditional management models separate the planning and the doing as far as possible. Managers are to plan, analyze, decide, and command. Workers should "check their brains at the door," be loyal and obedient, and, above all, to follow procedures and do what they are told. An engineer from Apple Computer called this model "flesh robots."[13] The model of "bosses" and "workers" is as much the core of Western machine-era business, as the model of "Lords" and "serfs" was for the feudal era.

Konosuka Matsushita, the founder of Matsushita Corporation, said that as long as the U.S. model for business is to put the thoughts of the managers into the hands of the workers, they (Japan) will win and we (the West) will lose. So far, he has been proven correct. As we move more into a service economy and place more strategic importance on high content, face-to-face transactions, the separatist mentality of pitting managers against workers is not only ill-advised, but fundamentally wrong.

Box 1-2

> **"My entire career was spent bringing firms into the Information Age. My group got the first IBM computer the Air Force bought in 1947. I spent 28 years as a consultant with Stanford Research Institute, working with clients in virtually all SIC codes and many nations. What helped clients the most, I think, was learning to develop multi-functional people and multi-disciplined project teams. It was hard for them to do..."**
>
> Dennis Finnigan, SRI International, Consultant Emeritus

"Process is all that matters — reengineering gone awry."

Consider trying to design a process that would allow driving through a major city and ensure that when you reached the 20th traffic light it would be green. That would be deemed impossible, and the request absurd. Yet it is seen as reasonable to ask things like "Give me a process to assure consistently successful high innovation products." The tacit assumption is that people don't matter, that they simply fit into reengineered processes.

The core of machine-age business was Taylor-era process. It allowed managers to increase their spans of control. It allowed workers without deep skills and managers without content knowledge or intuition to be effective. It allowed successful actions to be repeated without thought, and it allowed consistency of outputs.

There are two drawbacks to this thinking. The first is that a preoccupation with internal process diverts attention from creating external reality. This is the theme developed by Gary Hamel and C.K. Prahalad in their book, *Competing for the Future*. They argue that senior managers are spending less and less time on building a collective view of the company's future. Difficult issues are not confronted because they challenge the assumption that top management is in control, that they already have a clear picture of the future, and that they have better knowledge and a more accurate view than anyone else in the company.[14]

As a matter of recourse, senior managers like to focus on restructuring others. While this is sometimes legitimate and important, it has more to do with making companies lean-and-mean than with building competencies for the future. As Hamel and Prahalad put it, "Most layoffs at large companies have been the fault of managers who fell asleep at the wheel and missed the turnoff for the future."[15]

A second drawback of process thinking is that it works poorly or not at all for nonlinear problems. For linear systems — digging ditches or felling trees — traditional process is best. But imagine a nonlinear problem, such as an epidemic. Say that for every period, the disease doubles. On day one, you have two persons infected, on day two, four, on day three, eight, and so on. Developing a process-oriented solution to dispense limited vaccine and limit the epidemic to 500,000 is not as easy (a nonlinear task) a tasks like assembling coffee tables (linear tasks). Even if you were allowed to develop a new process specific to the epidemic, it would be difficult. In many cases, the sanctioned processes are embedded, inherited or adapted, making the problem even worse. AIDS, for example, is considered a civil rights, not a health, issue.

More confusing and disruptive is that linear systems — however simple or well-behaved — can become nonlinear, suddenly causing the methods that worked so well to fail totally. Usually this is caused by "outside" factors such as technology, innovation, market forces, or competitive actions. Peter Senge reports on MIT's "beer game" where teams of seasoned managers routinely and catastrophically fail to manage a simple beer distribution exercise when a single, minor, market nonlinearity is introduced.[16]

The best companies mix brilliance and consistency. Once a decade or so, some team at one of the truly excellent firms pulls off a heroic effort, successfully walking a high wire blindfolded while juggling razors, thus creating a legend. The rest of the time, these firms "simply" deliver superior products and develop loyal, happy customers. How do they do it? Partially through discipline, law and order, and attention to details — through process. Partially through passion, creativity, innovation, and risk taking. That takes knowing when to break the rules, the courage to do so, and the leadership, wisdom, and shared vision that guides when and how to take risks.

Culture is a good test. Some firms center on abundance and empowerment. They have process keepers but cherish stories that honor the human spirit. Stories extolling the low level team member who broke the rules to help a customer or meet a deadline. Stories celebrating the "eureka" that delivered the winning product, after many failures.

Other firms — this group is larger — focus on control, fear, and diminishment. Behind the corporate facade of slogans the employees privately tell how old Harry got his layoff notice by e-mail, how rewarded with Joe's good job was increased criticism, or how the CEO raged at Sally in some meeting. In these the bureaucrats, clerks, and sycophants rise to power.

"We can sustain our competitive advantage for a long time — a belief that may no longer be sustainable."

The pursuit of sustainable competitive advantage is central to the study of strategic management. The weathered book of standard procedures as reflected in mainstream management theories suggests that developing a sustainable gap against the closest competitor is the key to long term success and competitive survival. Differences in costs, prices, size, and resources provide the basis for creating competitive gaps.

The problem is that competitive advantages based on these traditional sources are becoming more difficult to sustain. Large scale firms, such as IBM, DEC, Wang, and General Motors, once considered dominant in the computer and automobile industries, have fallen prey to smaller, more nimble competitors. These former paragons of competitiveness now give the image of large beasts hopelessly mired in tar pits. Survival is often used as an indicator of effectiveness. Yet, among U.S. corporations, only 10% survive twenty years. Moreover, of those that survive twenty years, more than a fourth disappear during the next five years.[17]

The mortality rate is high for large firms, most of which have been desperately shedding jobs and cutting costs. The average life expectancy of a multinational is 40 to 50 years, and 1/3 of the 1970 Fortune 500 had vanished by 1993.[18] Even those that survive are troubled. The average real growth of the 1980 Fortune 200 was flat (–0.2% per year to be exact) over the period of 1980–1992. It was –3% if you strip out mergers and acquisitions.[19]

In *Hypercompetition*, Richard D'Aveni argues that competitive advantages are not sustainable because competitors can quickly imitate the leader, thereby eroding "first-mover advantages" and their attendant benefits.[20] A broader picture is presented by a leading Japanese management expert, Keniche Ohmae, who sees the disappearance of such advantages. Because it is difficult for any U.S., Japanese, or European firm to hold on to an advantage over time, it is more prudent for them to form global triadic alliances.[21] Unfortunately, with its institutions and precepts deeply rooted in distrust and adversarial law, the West has trouble with alliances in general and relationships in particular.

"Let's manage this crisis by controlling it — the hubris of traditional politics."

Since action is more valued than planning in the U.S., the way for a manager to get promoted is to effectively manage crises. Crisis management is seen to separate the men from the boys, the women from the girls, and the line from the staff. Is it any wonder that managers might allow or even create continual crises to hone or demonstrate this ability?

In such a world, footwork and persuasion are favored over integrity, homework, and substance. Stephen Covey called this the shift from "Character Ethic" leadership to "Personality Ethic" leadership.[22] The task of management is seen to reduce uncertainty in its external environment through proper forecasting of future states, through buffering the organizational core technology from random shocks, and through control of outcomes by means of short and long-term contracts. The management thrust in the U.S. is simplistic sound bites, quick-fix influence techniques, power strategies, communication skills, and positive attitudes.

The danger is that strategic managers, preoccupied with matters of control and prediction and "looking good," become trapped by the status quo. By failing to experiment, allow failure, and embrace change, these managers frequently miss the opportunities that allow success, and even survival.

In Eastern Europe, in Washington, at IBM, and everywhere, crisis management and central control are breaking down. By the time data is collected and analyzed, the situation will have changed so that even the "right" decision is now wrong. Workers now have more knowledge-depth than the bosses. If they don't, the organization is in deep trouble. The organizations that lurch from crisis to crisis are being outperformed by those who prevail with amazingly little fuss and bother. As Lao-tzu said millennia ago, "The great leader is one who causes the people to say, *We did it ourselves!* "

"What is made abroad can't be all that good — misplaced ethnocentrism in a global world."

For some time, it was believed that the decline in U.S. competitiveness was an inevitable consequence of post-war American policy to rebuild the economies of Europe and Japan. Unfortunately, the preeminence of the U.S. economy lulled industrialists and policy-makers into ethnocentricity and complacency. Learning from our competitors was deemed as hardly worth the effort; the focus remained at the U.S. domestic market. Consequently, while the rapid development of foreign economies had been occurring for decades, their new strength as trade partners was not acknowledged until we'd lost major markets. Complacency had taken its toll.

What underlies the friction between Japan and the United States is a difference in ideology. Depression-era U.S. economic theories and models assumed static markets and adversarial relationships between business, workers, and government. The U.S. government mostly views business as a zero sum game. It assumes that for every winner, there must be a loser. The key trade policies advocated by the U.S. are free trade, low tariffs, and open markets. On top of these, the U.S. government structure protects dying industries and encourages "pork-barrel" politics.

The preoccupation with domestic products leads to myopic assessment of international products. Japan's ascendancy as an economic power was not unnoticed by observers in the United States. Even so, reaction tended to be slow and guarded. Within the semiconductor industry, the distrust of products made in Japan was manifest. When the Japanese introduced their version of the 64k DRAM in 1984 at cheaper prices, Silicon Valley's first reaction was "they can't be that good." It was only after independent testing by Hewlett Packard proved the superior quality of Japanese chips that the U.S. industry took deep notice. By then, of course, we were far behind.

Reassessing Old Think Templates

We say the Machine Age models and templates are outdated, and for basic reasons. The Old Think business search is for the next thing that resembles the last thing that succeeded. What comes out of such Machine Age process is refinement of the past. In the Machine Age the firms structured for repetition (efficiently and with minimal thought) delivered better results than those who were not. These became the leaders in the Machine Age, and they had names like General Motors, RCA, U.S. Steel, and IBM. For decades IBM was always successful, but never at the leading edge.

The key to Machine Age success was running the "production line" faster, perhaps cheaper, and always with a minimum of top management attention. The key Machine Age assets were size — "fiscal and marketing clout" — not innovation. Therefore, Detroit clung to building traditional cars, IBM to "big iron" mainframes, AT&T to residential voice calls, and most of Hollywood to sequels and knock offs, rather than original works.

Machine Age managers seek refinement, not the first-of-type. The Machine Age was typified by a culture of competition based on bulk processing and "adequate" products. Production tended to be repetitive, much the same from day to day or year to year. Competing meant keeping product flowing, keeping costs down, and trying to improve quality. Since career advancement came from not being blamed for mistakes, new things were best avoided. New things, however promising, are so easily criticized.

There is an art to this sort of management, one discussed by decades of literature. It favors an environment free of surprises or glitches, one characterized by control and planning, one where customers have limited choices. It favors a hierarchy of workers and bosses. Machine Age workers should be, in the words of early efficiency expert Frederick W. Taylor, "no smarter than an ox." Because bulk processing is repetitive, it allows constant improvement. And so, this is a world that favors internal focus, hierarchy, planning, standard procedure, and controls. Above all, it is a world of optimization.

Still, all the optimization in the world could not help phonograph records compete with compact disks, propeller driven airliners compete with jets, or movie cameras compete with VCRs. All the process optimization in the world couldn't help IBM's mainframes or DEC's minicomputers compete with microprocessor-based distributed computing. Nor could it help Apple's 1984 Mac compete in the PC-based "WinTel" world of the 90s.

Unlearning Old Think

There is a Zen-Buddhist story about a Western man who was considered a leading thinker of his time. He had several degrees in philosophy and wanted to travel to the East for an encounter with a Zen master. He traveled thousands of miles for a meeting with the master, and then waited many days for a private interview.[23]

When the interview was granted, he finally sat at the table before the master. Tea was brought into the room. As a servant placed cups before them, the Western man began to regale the master with his accomplishments, his credentials, and the many books he had read and written.

The master was silent as the other talked on and on about the knowledge he had acquired, explaining that while he wanted to be the master's student; his teacher should realize how much he already knew. Time passed as the visitor warmed to his story, proudly and eloquently citing his many achievements.

Finally, the Zen master carefully picked up the teapot and began to pour tea. Abruptly, the man stopped in the middle of his monologue. The master had filled his guest's cup, but he continued to pour. The tea overflowed the lip, filled the saucer, ran across the table and dripped over the edge, finally splashing on the visitor's feet. "You're pouring tea everywhere," the man exclaimed incredulously.

To which the Zen master replied, "You are like this teacup. It is full, so it cannot contain anything new. You are so full of yourself, you have no room for anything else."

Why Old Think Prevails

Because top managers adamantly cling to weathered beliefs and perceptions, few things change. In fact, the adage describing such behavior is: "If it ain't *really* broke, don't fix it." When their jobs are at risk, few dare to point out that the Emperor lacks clothes, so change is slow. And because organizations tend to first respond to change in the business environment by adjusting standard operating procedures, most changes are cosmetic. The four most common scripts used to justify an inability to change are *inertia, denial, scapegoating,* and *calibration.*

Defensive routines: inertia

Inertia, or resistance-to-change, is rooted in how organizations learn. Two leading scholars, William Starbuck and Richard Nystrom, have spent most of their careers examining organizations in crisis. Interestingly, they argue that organizations fail because they are blinded by their past successes.[24]

Successful firms weather stormy transitions and precipitous environmental "shocks" by developing a repertoire of finely tuned adaptive routines. Over time, these routines, termed "standardized operating procedures," become institutionalized within their corporate structures and processes. Such routines, once historically circumscribed to specific actions, become more general responses to demands on these organizations. In stable environments, routines provide the comfort zones, the buffers that accommodate change and lead to consistency. In highly unstable environments, these routines often become rigid structures, prisons that impede innovative solutions.

There are many manifestations of inertia, but none as articulate as the phrase, "Yes, but..." Peter Senge, an academician who has examined learning organizations, described blindness and rigidities as limiting learning within organizations.[25] All too frequently, organizations that fail to learn are those that find reasons not to do anything. Another unintended consequence of inertia is "learned helplessness." This behavior occurs when a prospective change-agent within an organization finds himself surrounded by naysayers and obstacles. Soon, this person will shy away from attempts to drive change, claiming that these would only be a waste of time. The encased learning of Old Think has to be discarded, not adjusted, for new learning to occur.

Denial & Scapegoating

For many people, a common form of defensive behavior is denial. "Other companies may have problems, but not us." Other related behaviors include the evasion of conflict by procrastinating, or constructing wishful rationalizations, or remaining intentionally selective about what information to process.

Denial also comes close to scapegoating — blaming another party for one's misfortune. Various social scientists have studied the process of attribution, or how one assigns causes to particular actions. Internal attribution is defined as assigning cause to oneself, or to items under one's control. External attribution refers to the assignment of cause to factors outside one's control. In tests of these processes, scientists have discerned patterns that suggest that successful outcomes tend to be attributed internally. ("Our sales increased because of our hard work.") The reverse occurred with unsuccessful outcomes. These are attributed externally. ("We did not succeed because the government offered us no real protection.") [26]

A popular scapegoat for the competitiveness problems of U.S. firms is the Japanese. Clearly, some misgivings about Japanese industrial practices are justified. Still, feverish sentiments towards protectionism seem to peak when U.S. firms are not doing well. At the height of Japan-bashing many years ago, it was suggested that the Japanese succeeded because they worked too many hours a day, and that they did not have as much leisure time as the Americans or the Europeans. The ensuing suggestion that they relax and play more golf struck many as outlandish and ethnocentric.

Calibration

Central to Old Think is the hope that the old beliefs will prevail, given enough time. On the surface, there appears to be no shortage of recommendations for corporate renewal. Without much precedent to guide them, managers have engaged a wide set of actions that include downsizing activities, the empowerment of employees, the building of corporate renewal, and the formulation of transnational strategies. Managers are watching hierarchies fade away, and finding the traditional distinctions between functional units to be much less compelling. Economic changes are forcing organizations to become leaner, less bureaucratic, and more entrepreneurial. As traditional command-and-control systems erode, there is the need for newer models of leadership.

On the surface, it is tempting to view these new paradigms as passing fads, and to hope that, consequently, the traditional theories, once better understood and correctly applied in context, will still provide the right answers. Many managers hope that incremental refinements of traditional theories will be adequate. They hope that firms in transition are merely settling into a "new" state of equilibrium; downsizing activities will eventually result in the optimal "rightsizing" levels; leaner and more bureaucratic firms are needed manifestations of much-delayed discipline; and emerging models of internationalization will eventually reflect firm-derived sources of competitive advantage. Others just keep their heads down and make sure that their golden parachutes are in place.

New ways of learning are needed. Because Machine Age thinking is regularly reinforced in mainstream theories, they also get in the way of learning. As a consequence, the proverbial teacup is too full to accommodate new ideas. Learning must start with the unlearning of old ways. A new premise is that people are assets, not expenses. Management leverage must move from financial manipulations to the timely and profitable creation of knowledge. The structure of the corporation will change — from hierarchical, control-oriented structures to protean structures that will flex and change with times. Control-centric, procedural management must and will shift to empowerment and leadership.

The Old Think firm's own internal processes and biases dictate more of what it does than do the merits of new products or the needs of the external environment. The key premise, usually unstated, behind all this is that doing more of what worked in the past is the key to success in the future. The new milieu, as described in the next section, is very different. Management is not production oriented, but services oriented. Hierarchies flatten, not because computers and process can flatten the organization (as re-engineering advocates would tell us), and not because sensitive managers suddenly favor democracy (though some may). They flatten because, to be effective, the deliverers of the next-thing-for-the-company need to be organized like commando units, in small teams that are empowered to meet and to make decisions with the customers at virtually all levels of the organization.[27]

New Think: From Machine Age to Information Age Imperatives

Can you accept in your heart and mind the crucial, critical, central fact that business and society have now entered a totally new age? It has. Some call it the post-Capitalist or post-Industrial era, others simply call it the Information Age.[28] This era is fast displacing the Machine Age — a change that demarcates a significant historical divide. Some experts believe that the human race now faces more change than at any time in all of recorded history.

Box 1-3

> **"All understanding starts with our not accepting the world as it appears."**
>
> Alan C. Kay, Apple Fellow (during their best years)

Borne out of Newtonian physics and the Industrial Revolution, the Machine Age stressed the importance of tangible, physical resources as the source of wealth. It became the basis of rules that emphasized physical specialization, natural resources, factor inputs, and crude energy sources. The traditional language of business (financial accounting) and the assumptions related to wealth creation (value-added accounting) reflect these rules.

Financial transactions are used to categorize physical activities and to evaluate wealth accumulation on the basis of factor inputs. Managers pay more attention to numbers than technology, products, or customers. People are expenses, not assets. Firms are increasingly run by accountants and clerks. Management structures are hierarchical, functional silos, not integrated wholes.

Examples of where "management-by-accountants" has been expanded beyond sane limits are abundant. Consider HMOs that advertise they "involve doctors in the decision making practice for their patients' care," as if that was a remarkably magnanimous act. A jet engine firm was purchased by a company whose prior experience was in supermarkets. The new owner quickly computed profitability attributed to each department by the floor space occupied. On that basis, it abolished the "least profitable" departments, metallurgy and fluid dynamics, the very basis of jet engine manufacture.[29]

Mixing government with accountants produces strange, even scary, results. The Megastate proposes selling off national treasures — from the Patent Office, a quasi-judicial function that safeguards a Constitutional right, to the National Parks — to generate revenue. There was 1997 Senate testimony that the IRS was illegally shaking-down taxpayers based on how much money it could cheaply collect, versus how much the poor wretches actually owed. Still, the places where this accounting mentality has prevailed (AMTRAK, the post office, the forest service, broadcast licenses) are hardly encouraging. The IRS itself has failed several GAO audits and can't account for 64% of its congressional appropriation.[30]

Conversely, the new precepts have a different, non-financially centric focus. The Information Age has its origins in quantum physics and upholds the power of ideas and technologies over physical matter and resources.[31] As such, technology and ideas free people from the routine of Machine Age process. Information allows people to use their knowledge, intuition, and training. The conduit of change is technology, particularly the advances in microelectronics. This has allowed exponential progress in computers and communications, now spotlighted by the Internet phenomena.

In *Microcosm*, George Guilder amplifies this argument: "Wealth comes not from the rulers of slave labor but to the liberators of human creativity, not to the conquerors of land but to the emancipa-tion of the mind."[32] Information allows people to use their knowledge, intuition, and training. The physical attributes of the product are not as primary as the value of services co-produced by both the buyer and the seller. Accounting systems based on factor accumulation are inadequate; more sophisticated measures of service transactions are needed.

Knowledge-based competition is fundamentally different, and for the most basic reasons. Its dynamics, basis of competitive advantage, enabling technologies, and economics are all-different. Today we face the business equivalent of the discontinuity between Newtonian Physics and Quantum Physics. Newtonian Science produces large machines that can be studied by disassembly into component parts. Quantum science produces magical constructs of the mind that cannot be fathomed at all without holistic understanding. Rather than diminishing economic returns, increasing returns dominate.

It took several generations for science to adapt to its new world, but business won't be given that long. Today's global markets tend to be "winner take all," so most of the losers won't be around to adapt.[33] The goal in high prosperity industries is the search for "the next big thing," and the expectation is that it will not resemble the last big thing much, if at all. Microsoft's dominance and core competencies in desktop computing may or may not be relevant to the world of the Internet, depending on what choices made by Bill Gates and his team.

Box 1-4

> **"In today's global Information Age economy, anything that is standard and routine quickly falls to third-world wages and razor-thin margins."**
>
> John D. Trudel, 1995

If the purpose of business is creating customers and growth, then the new knowledge-based industries are the high ground. Some firms and some nations take that view. Many smart nations and firms are only too happy to outsource to Old Think companies, or to buy them up at fire sale prices. Sun Tzu counseled millennia ago to always allow your enemy a path of retreat.[34] The path of retreat is to be "cheap" and "adequate." It is an open path, but often a trail of tears. Generic products soon become commodities with thin margins. Because of chronic global over-capacity, they are often sold below their original cost or produced in third-world sweatshops.

Successful Information Age leaders target new, high value-added, high prosperity business opportunities. These seek to create a prosperous future by making past competitors irrelevant. Still, this rich opportunity comes with its own challenges. In knowledge-based business, missing the next product wave is disastrous. Missing the next industry is worse, and all too common. Few leaders made successful transitions from vacuum tubes to transistors, or from transistors to integrated circuits.[35] Few leading computer firms made successful transitions from mainframes to mini-computers, from minis to workstations, or from workstations to PCs.

There are existence proofs — Hewlett Packard is a good one — of organizations "built to last," but these are few and far between in the West.[36] Why do the past leaders so often lose in today's markets? Compared to those who topple them, the losers start with almost limitless resources in the form of money, talented workers, distribution, and loyal customers. How can they lose with such consistency, when traditional management wisdom — at least from the Olympian perspective of the mega-consulting firms and B-schools — would say that they should win most or all of the time?

Regardless of "why," the reality is that established firms do lose consistently to smaller, weaker upstarts who exploit new technology and knowledge to raise the bar and negate the strengths of the former leaders. The upstarts focus on technology and opportunity, both internally and externally. They seek high value innovation and to quickly match technology and core competency to emerging need. The upstart firms run fast and "eat their young" without remorse. They joyously cannibalize their old products, not to repeat the last success but to raise and reinforce competitive barriers.

The firms that are winning today are vastly different organisms from those that prevailed in the past. They are smart, quick, and resource efficient. In every market and in every industry these new organizational life forms are winning out over ponderous functional hierarchies. Marvin Patterson from Hewlett Packard uses the analogy to a Cheetah chasing its prey.[37] The successful predators in the fast-cycle markets of the 90s intuitively know where to go and how to get there quickly and efficiently. Functional organizations, while good for some things, can't do that.

The New Competitive Landscape: The Engines of Prosperity

While most executives recognize that competitive patterns in their external business environment have changed, a deep understanding about what has happened, or how and why it has occurred, is still lacking. The problem is that our way of thinking about the environment has not changed.

Evolutionary biologists have discovered that the world follows "punctuated equilibrium."[38] Species may flourish for millennia until there is a discontinuity that makes them unable to compete, after which they quickly die off or decline. The same is true in business and economics, but the major discontinuities are so infrequent that centuries may pass between them. The shift from the Machine Age to the Information Age is such a discontinuity, and one that will have more impact than any other event in human history. By comparison, the industrial revolution of 400 years ago was only a minor ripple.

Such changes in the external environment are not well understood, much less their causes or their significance. As with an earthquake, slow pressures can cause cataclysmic change when the system suddenly snaps to a new equilibrium point. So it is in many ways with U.S. competitiveness. A feature of the new competitive landscape is the rapid diffusion of technology across the world. An economist, Joseph Schumpeter, once argued that the driving force of any economy is the creation or innovation of new products and processes, which in their success, destroy the prevailing technology.[39] Each wave of technology would usher in new investments and provide new jobs in the new technology.

Later models focused on the dynamics of technology's life cycle. Pundits argued that technologies followed the path from introduction and growth, to maturation and decline. Yet it is this model's assumed inevitable march to maturity and equilibrium that presents its greatest weakness.[40] It does not match Information Age reality. It cannot explain industries that seemingly never reach maturity (consumer electronics) or decline (management services), nor can it explain industries that undergo "de-maturity" (textiles). In addition, advances in microelectronics, in particular, have created new industries (microcomputers), while destroying others (mainframes). Some industries have shown a seeming ability to be renewed with each new product or technology cycle.

Evidently, the new mind set for operating in this environment draws its foundations from a radically different basis than what characterized the first two industrial revolutions. We argue that these new foundations are already in place. These are grounded in the movement from Newtonian to Quantum Thinking, from Cartesian Division to Holistic Patterns, from Equilibrium to Ordered Chaos, and the New Growth Model theory of increasing economic returns.

From Newtonian to Quantum Thinking

Conventional economics has tended to portray the economy as akin to a large Newtonian system, with a unique equilibrium point preordained by patterns of mineral resources, geography, population, consumer tastes and technological possibilities.[41] In this view, perturbations or temporary jolts — such as the oil shock of 1973 and the stock market of 1987 — are quickly negated by the opposing forces they elicit. In view of future technological possibilities, one should be able, at least theoretically, to forecast accurately the path of the economy as a smoothly shifting solution to the analytical equations governing prices and quantities of goods. History, from this vantage point, merely delivers the economy to its inevitable equilibrium.

Positive-feedback economics, on the other hand, finds its parallels in modern nonlinear physics.[42] Ferromagnetic materials, spin glasses, solid state lasers, and other physical system that consist of mutually reinforcing elements show the same properties as the economic examples. They "phase lock" into one of many possible configurations: small perturbations at critical times influence which outcome is selected. The chosen outcome may have higher energy (that is, be less favorable) than other possible end-states.

This kind of economics also finds parallels in the evolutionary theory of punctuated equilibrium. Indeed, bio-economics is now a recognized discipline. Small events often "average out," but not always. Once in a while they become significant in tilting parts of the economy that result in new structures and patterns. These, in turn, are preserved and built on in a fresh layer of development. Accordingly, economies that may be initially identical may eventually diverge with significant increasing returns. Divergence limits forecasting. Futures are not so much predicted, as they are created. Changes follow a gradual accumulation of stress, which a system resists until it reaches its breaking point, or until a triggering event precipitates discontinuous change. This is why punctuated equilibrium is characterized as changes that occur in large leaps.[43]

From Cartesian Division to Holistic Patterns

For most of the modern era, the metaphor of the *Machine* provided the paradigm for understanding the world. Rooted in 17th Century science (Newtonian Physics), this metaphor conceived order as a Cartesian whole, for which motion was determined by fundamental laws.[44] Social scientists were quick to adopt this paradigm. One implication of this was the separation of the observer from the observed world. One characteristic feature was: *The Whole = The Sum of Its Parts.* Understanding the parts through analysis was essential in defining the Whole.[45]

With the advent of quantum physics (and evolutionary biological theories), a different world-view emerged. Reality was not observed per se, but patterns of relationships defined reality. Parts derive their meaning from the whole; the whole derives its quality from the nature of its parts. Things matter, but so do the white spaces between things. It was Gregory Bateson who observed that reality is the pattern that connects (or one that creates meaning for people).[46] Thus, reality transcends the limitations of a Cartesian plane, i.e. an x and y axis, and is based on the interconnections of parts.

Within quantum physics, theoreticians observe and discern these patterns. In their view, light can be proven to be either a particle or a wave, depending *entirely* on the observer's viewpoint. The future is no longer so much a matter of understanding nature with the end in view of controlling it. We are a part of nature ourselves. In the words of Henry Miller; "It is for us to put ourselves in unison with order." Deo Strempkof, a psychologist from South Africa, once observed that: "In nature, all the colors blend."[47] Or, as Information Age leaders are starting to say, "We create our own future."

Traditional management theories are built on analysis, functionalism, and control. The old Taylorism would break down parts of a job in order to evaluate its overall efficiency. Traditional MBA and management education programs are based on the so-called functional disciplines. And, management control, through numbers or behavior, represents the kernel of accounting and budgeting systems. The world is changing in ways that no longer facilitate this paradigm. A shift to one that promotes the discovery of new patterns of meaning is needed.

From Equilibrium to Ordered Chaos

A related anchor of mainstream management theories is *equilibrium*. It is widely acknowledged that successful businesses operate as close as possible to conditions of equilibrium. The task of management is to reduce uncertainty in its external environment, through proper forecasting of future states, through buffering the organizational core technology from random shocks, and through control of possible uncertain outcomes by means of short and long-term contracts.[48]

Management theories are generally developed from equilibrium and order. While there is reference to disturbances ("turbulence in the environment"), eventually the system converges into an equilibrium state. In Michael Porter's now classic strategic planning framework, for example, organizations are reassured of order when developing specific generic strategies to protect themselves from the competitive forces within their industry.[49] Proper implementation occurs when strategies are co-aligned with structure and processes. All in all, disequilibrium is conceived as a transitory state before equilibrium seeking solutions that brings the system into balance.

The quiet revolution, led by a group of academic mavericks, such as Brian Arthur, Murray Gell-Mann and Kenneth Arrow, has challenged the kind of linear, reductionistic thinking that has dominated science since Newton.[50] Using novel ideas about interconnectedness, coevolution, chaos, and sharing, they have been examining unpredictability that has not been explained by conventional theories. These theories describe many ways in which order arises out of apparent chaos in physical systems, as with chemical solutions. They address the dynamics of a self-reinforcing system, i.e., one which repeatedly builds on the results of its own interactions to achieve more richly ordered complexity, rather than repeatedly damping its own effects to remain simple and straightforward. Each self-reinforcing system is believed to contain greater possibilities. In effect, the system amplifies itself and expands. Along with expansion comes unpredictability and novelty, and chances for something new emerge into the world.

From Exogenous to Endogenous Conceptions of Technology

This new impact of technology is the subject matter of Paul Romer, a ground-breaking economist already touted as "arguably the most influential theorist of the 1980s."[51] Romer is the leading proponent of New Growth Theory, a branch of economics that attempts to identify the underlying causes of growth. While traditional economic theories emphasize two factors of production (capital and labor) Romer adds a third — technology.

Three arguments about technology highlight Romer's work. First, technology is "endogenous," or within the system. Although any given technological breakthrough may appear random, technology overall increases in proportion to the resources devoted to it. Second, technology can raise returns on investment. Technological advances will create opportunities, leading to increasing rather than decreasing returns to scale. And, thirdly, investments can make technology more valuable and technology can make investments more valuable as well. This third argument signals a new role for monopolies and possible, large conglomerates, as these are the organizations that can afford sizable investments.

Despite these advantages, the leader's market power is tempered by the rapid pace of innovation, which encourages leapfrogging strategies and even radically new entrants. Some critics wonder why market leaders fail to sustain their edge when they enjoy monopolistic power from lock-ins and bandwagon effects.[52] Unfortunately, the leaders' preoccupation with sustaining their edge can also lead to myopia in recognizing new technologies or incipient strategies of late entrants.

If a market leader becomes complacent, it can be "blind-sided" by a new entrant. It is now commonly acknowledged that IBM's preoccupation with mainframes precluded proper market attention to personal computers. Seymour Cray's fascination with supercomputers led him to disregard microprocessors, small machines, and emerging software markets. The uncertainties of new technology makes leaders tend to paranoia.[53]

While Microsoft appears to have a sizable lead in PC software, it is very concerned about the rapid emergence of Netscape and the Internet that can change the competitive rules-of-the-game. Similarly, Intel's lead in microprocessors is assured only to the extent that people continue to use PCs and not other platforms. All in all, competing in this environment involves an understanding of the dynamic context (engines of prosperity) and game-theoretic strategic responses (disruption strategies by the follower).

Taken collectively, these works provide the foundations for a competitive landscape based on technology-generated industries. Our premise is that the world has shifted in significant ways that fundamentally challenge traditional theories and assumptions about management. Three shifts are particularly manifested. First, the information age has ushered in newer conceptions of time that erode advantages initially derived from lead times and slack resources. A few leading-edge firms are using time, synchronicity, and organizational learning as competitive weapons.

Second, traditional management theories, oriented to capital-intensive industries, still fail to acknowledge this. Partly in response to information technology, newer organizational structures consist of networks, hallowed relationships, and contractual alliances. Traditional management theories emphasize variations of vertical, separate, and functional structures.

Third, new competitors, such as the Japanese and newly industrializing countries, have successfully forged partnerships among government institutions, labor, and business firms. Our review of traditional management theories, directed at the experiences of U.S. firms, indicates that they still de-emphasize the important effects of institutions, multidisciplinary teams, and infrastructure.

> Definition: **Engines of Prosperity — industry and organizationally based conditions that generate non-linear patterns of growth and renewal.** See Chapters 5–7.

Taken collectively, this new environment can be characterized in terms of basic patterns, which we call the engines of prosperity. These patterns are driven by modern technology that has its origins in quantum physics, its embodiment in the microchip, and its exemplary product in the computer. Patterns of growth and renewal, based on emerging theories, exhibit the following:

- steep learning curves,
- demand amplification, or increasing returns,
- technology generators, bandwagons, competitive imitation, and lock-in,
- amplifying or disruptive competitive behavior, and
- the unbundling of the firm's value chain.

While these characteristics are cited in books, oftentimes as isolated examples of how to build competitive advantage, their treatment is fragmented and arbitrary. In this book, the new technological environment becomes the relevant context for examining these characteristics in a holistic fashion. We'll say more later about the implications of these engines for competing in a fast-paced environment that is driven by continuously accelerating technology. For now, we make four points:

- There are severe limitations to static, linear models as applied to a dynamic (non-linear or quantum) world. Often the old models give the wrong results.
- There is an increased emphasis on speed, on shortening the already shortened life cycle. When trading off time for money, you should often pay more or suffer business "losses" to be early.
- Reliance on historical accounting methods proves illusionary in measuring true costs and opportunities in the Information Age. Traditional accounting can precisely measure the past costs and benefits of physical assets. It does not do well at measuring knowledge assets (people, ideas, technology, and intellectual property), dealing with uncertainty, or forecasting future revenues.
- There needs to be a balance between the requirements for sustaining momentum in contexts where the engines of prosperity are in operation, and for nurturing radically new technologies that can upset growth trajectories and create new markets.

Competing in the New Environment

The Management Challenge

Perhaps the most critical challenge facing today's organizations is how they can continuously renew themselves in the midst of fast-changing conditions. It is widely acknowledged that for firms to remain effective, they must capitalize on opportunities in their external environments in ways that leverage their internal capabilities and minimize their weaknesses. Even so, the selection of a good strategy by itself does not guarantee success.

Today, leading-edge theory and progressive management practices focus simultaneously on strategic analysis, organizational structure, timely content knowledge, and management processes, adjusting all four interventions to achieve coordinated patterns in the race against quick competitors and a rapidly-changing dynamic environment. The context in which this coordination takes place is a fast-changing environment in which competitive rules are not likely to be defined with clarity or confidence beforehand, and where powerful coalitions within the organization may tend to be parochial about their spheres of influence.

Organizational renewal in this new environment requires adaptation skills that are based on the engines of prosperity. We argue that firms need to develop better sensitivities to their environments, probably achieved by unlearning, purposeful discovery, and organizational learning. Recognizing that competitive advantages are not sustainable, they also need to plan for destroying these advantages in order to allow new ones to flourish. Finally, they can develop learning infrastructures that accommodate any corporate strategy, rather than aligning particular (generally functionally specific and rigid) structures to strategies, as suggested by conventional theories.

The Leadership Challenge

The pursuit of sustainable competitive advantage is the goal of strategy. By creating a sustainable gap, organizations can effectively exploit their advantages and achieve above-average performance. The conventional theory is that advantage can be sustained by engaging in cost leadership or differentiation strategies in broad or narrow target segments. However, even successful Japanese firms now realize that these advantages cannot be sustained over time.

In contrast to conventional approaches to sustaining advantage, the strategic challenge in the new environment is to balance learning with unlearning. In his book, *The Fifth Discipline*, Peter Senge compares an organization to a learning organism, with concomitant learning disabilities. Using a systemic framework, Senge has identified recurring patterns of disabilities within organizations that impede their learning, or even more adverse, predispose them to perceive opportunities and threats in a rather narrow and myopic view. Unless such disabilities are correctly diagnosed and corrected, the organization is impaired from effectively adapting to the needs of its constituents in its external environment.[54]

Learning is also linked to timing and risk-taking. Organizations fall prey to errors in timing by entering a market prematurely, or by entering it too late. The timing of market entry and know-how can help the development of proprietary assets. First mover advantages place pressure on the follower to enter the market at a higher cost burden. The risk, however, is that if imitation occurs too quickly, then the leader might not have sufficient time to recoup its investment.

More subtly, both the timing and levels of investment are crucial. Early resource has more leverage, but failure can be caused by either too much or too little resource. Knowing when, where, and how much to invest is a difficult art, as is knowing when to stop. The balancing act between being early to market and having the right product requires exceptional talent, courage, and vision. Intuitive decisions are required, so these skills must be developed.

The Institutional Challenge

What makes international competition distinct from a purely domestic one is the nature and source of competitive advantage. We argue that competitive strategies of firms, operating in both traditional capital intensive and knowledge-based high technology industries, are inextricably linked to the larger institutional context in which these strategies are developed. The success of Japanese high technology firms is a testimonial that directed government intervention and supporting national institutions can play a key role in successful corporate strategies. Firms can derive advantages from their embedded institutional environments, or what James Moore calls the "business ecosystem."

In his provocative best seller, *The Death of Competition*, Moore cautions on the fallacy of viewing the corporation as a "firm."[55] Even a well-managed firm with an excellent product, he argues, can fail if its extended suppliers and customers are poor. He argues that the proper unit of analysis is the "ecosystem," defined as the vast set of relationships between the firm and its environment.

Moore provides the astute reminder that firms are part of a larger entity, and his analysis can be carried further. It is possible that a firm's competitiveness is tempered by what actions are made within the ecosystem. The Internet is a good case in point. As customers explore this system and continuously find new applications and opportunities, there is the counterbalancing force of government. As usage grows exponentially, so does governments' desire to control. Note the attempts by Germany to control content, the phone companies' lobbying to raise line charges, President Bill Clinton's unyielding "clipper chip" initiatives, and the quest of government at all levels — national, state, and local — to tax or regulate it.

Unfortunately, Megastate bureaucracies dislike citizen self-reliance and abhor indirect roles. All too often, they develop strong symbiotic relationships with the very problems that justify their existence. These vicious cycles are a bureaucrats' dream. The more that is spent, the worse the problem, the more that must be spent, etc.

The worst consequence may be that Megastate bureaucracies, whether in Washington or Eastern Europe, tend to lock companies into Old Think. Corporate bureaucracies are seen as a safe way for firms to interface with the bureaucracies that have dominion over them, many of which have the power to enact and enforce law. Perhaps Information Age firms will need buffer zones ("Departments of bureaucracy?") to connect to the Megastate.

Still, the economies that are doing best at trade all exploit versions of the Japanese model of government economic partnership with business to create high wage jobs and national competitive advantage. Clearly, national economic competitiveness is not a priority for the U.S. government. For example, it is perfectly legal for General Motors and Toyota to partner, but not for General Motors and Ford.[56] Employment law is used to make companies hire the less qualified, including convicted murderers and felons. Information Age business needs skilled workers, but one must go through a dozen or so priorities in public K-12 schools before you get to any goals that involve education. National programs like Outcome-Based Education (OBE) exacerbate matters by shifting metrics from academic achievement to social conformance.[57]

"We're from the government, and we're here to help you," is true in many Pacific Rim nations, but a common joke in the U.S. Therefore, despite risks, because respect for the government is low, because time is crucial, and because the cost of technology is increasing, banding together by private partners is a major and growing trend.

It is hardly surprising that many executives have dropped into a foxhole "just keep 'em off my back" mind-set about government. The institutional challenge therefore often falls to the private sector, which is ill equipped to respond. One common solution is to move jobs to locations with better infrastructure. Safe harbor nations, like Singapore, are booming.

Scope and Organization

Engines of Prosperity is premised on a significant idea with far-reaching consequences: the advent of the new information age will impel new ways of thinking based on a thorough understanding of emerging patterns of growth and renewal. The book is organized into three major parts. Part One — Environments and Models — examines the nature of competitive environments faced by contemporary firms, and how changes are breaking down traditional conceptions of industries and management structures and methods. Chapter 1 presents the core arguments, Chapter 2 describes the emerging knowledge-based economy, Chapter 3 examines the dawning of global markets and industries, and Chapter 4 presents the new technology.

Part Two — Examples, Rivals, and Philosophies — starts with specific cases of head-to-head conflict in dynamo markets. We first examine the potential and limitations of current strategy models, and the foundations of the new economic order (Chapter 5). We then present extended examples of how the engines of prosperity have changed the essence of competition in a number of industries, with specific attention to videocassettes and semiconductors (Chapter 6). Then, we discuss the specific implications of a new environment and their orientations (Chapter 7).

Finally in Part Three — Actions and Attitudes — we synthesize various themes throughout the book into a structured argument for restoring the competitiveness of U.S. firms. Policy options to restore Western organizational, strategic, and institutional competitiveness are presented in Chapters 8, 9, and 10.

Chapter 2

The Emerging Knowledge-Based Economy

Key Themes at a Glance

Business has entered a new age. Some call it the post-Capitalist or the post-Industrial era, others call it the Information Age. The emergence of a *knowledge-based economy* is the first of three major trends that is faced by the 21st century manager. The other two — discussed in the next chapters — are the *dawning of global markets* and *advances in new technologies.*

We contrast the Information Age to its predecessor — the Machine Age. Machine Age models and paradigms are no longer helpful. They get in the way of progress. New ways of learning are needed. The new learning must start with unlearning the old ways.

The Machine Age "language of business" (traditional accounting) is poorly suited to managing knowledge. People are assets, not expenses. Management leverage has moved from the manipulation of financial abstractions to the timely and profitable creation of knowledge.

World power and national prosperity will now be determined by ability to compete in the global markets of the Information Age. This dooms the Megastate, which, though almost universal, has not worked well anywhere. It also dooms most industrial age firms.

The essence and structure of corporations will change. Companies will organize around information, not things.

> **Structures will be "Protean," they will flex, change, and evolve with time.**
>
> **Control-centric, procedural management will shift to empowerment and leadership. The new 21st Century Taylorism — where standard procedures, processes, and workflows are set by the workers — is a little better than the old form, where these were set by "experts."**
>
> **There will be smart, prosperous nations and ignorant, impoverished, backwaters, with a growing gap between them. Only knowledge and productivity will allow prosperity. This necessitates retooling with new infrastructure, like broadly deployed, inexpensive, secure fiber.**

Never before in the history of business have the issues of "change" and "turbulence" been so salient and unforgiving. Advances in technology have ruptured industry boundaries and overthrown standard management practice. Simultaneous upheavals in politics, technology, and economics have created a new order. The collapse of communism and the liberalization of centralized command economies are driving global commerce and international investment.

The continuous developments in communications and information technology have forged stronger links between nations, creating extricate networks between nations, companies, and peoples. Paradoxically, this increases opportunities for both cooperation and conflict. Where once nations fought over trade routes and raw materials, in the future, they will battle for intellectual property and the right to build and globally market unique products.

The Advent of the Information Age

Fascination with knowledge-based society had long preoccupied Western scholars. Indeed, the U.S. Constitution (Article one, section 8) marked the world's first right of citizens to own intellectual property. Our founding fathers saw knowledge as the key to building a prosperous society. The original Patent Commission had George

Washington as CEO, and included the Attorney-General and the Secretaries of State and War.

Combining individual reward with technological advancement was a winning combination. President Harrison said so at the Centennial of the U.S. patent system. "It distinctly marked, I think, a great step in the progress of civilization when the law took notice of property in the fruit of the mind."

From the 1760s to the 1830s, steam engines, textile mills, and the Enlightenment produced the Industrial Revolution. The Civil War saw a renewal of military technology, which had languished for over three centuries.[1] The years 1880 to 1930 were shaped by the spread of electric power, broadcast, telephones, mass production, and democracy. On the eve of the 21st century, we are witnessing the end of the Cold War, the emergence of economic conflict and integration, and the pervasive impact of new technology. The true impact of the knowledge revolution has just started.

Alvin Toffler, Peter Drucker, James Brian Quinn, and John Handy have, in their own ways, heralded the emergence of a new economy based on information and knowledge-creation.[2] In fact, Drucker has forcefully and persuasively argued that knowledge is not just another resource that adds to the traditional labor, land, and capital — but that it is the only meaningful resource today.[3]

Profound changes start quietly. Nevertheless, fundamental drivers that define a particular era drive these changes. Much of the competencies developed in the 17th to the 19th century were land-based, or, more generally, factor-based. The theory of comparative advantage is still the centerpiece of every course on the theory of international trade. It is the basis for estimating a country's competitive advantage in the production of goods and services for international trade.

Initially formulated by economist David Ricardo, it postulates that comparative advantage is based on land, labor, and capital, and that nations should produce those goods for which they have the greatest relative advantage.[4] While the premises of the theory tend to be crude and oversimplified, it had value in providing the original context and logic for international trade.

The 19th–20th century ushered in a refinement, known as Heckscher-Ohlin (after the two economists who developed it) or, more commonly, as factor endowment theory.[5] The theory postulates that nations will produce and export products that use large amounts of production factors that they have in abundance, and that they will import products requiring large amounts of production forces that are scarce in their country. This theory helps explain why China, endowed with abundant cheap labor, will produce labor-intensive goods, while the United States, which has more capital, will specialize in capital-intensive goods.[6]

Note carefully that competencies, in this context, are all grounded in production efficiency. The Machine Age was about mass, routine production of adequate goods. Machine Age templates emphasize process, repetition, and production. Reliability and quality were grafted on later, though with some difficulty and only after "mavericks" like Japan had captured lucrative markets. With such a production-centric mind-set, it is hardly surprising that both scholars and practitioners have looked to economies of scale in determining the competitive profile of a country. These models have worked for decades, but they are no longer adequate or even particularly relevant.

Information Age technology has shattered the old models. The new competitive differential is the creation and application of technology, a form of knowledge. In cases where innovation is important and where the resources of any country are not endowed, but created, then new issues are raised. Governments of Japan and Korea, among others, have made industrial policy a major part of their planning process and national competitiveness a priority.[7] Some have moved past factor endowment to newer models like Kodama's "technology fusion." Even in production-centric cases, the key issues are now knowledge. For example, Korea's developing electronics industry has typically paid more to Japan for intellectual property rights than it spent in total on components and labor.[8]

Choices for how to react to the new world economy are diverse. Some nations, especially those on the Pacific Rim, are aggressively building new infrastructure and competencies to ensure future high wage jobs and prosperity, often through "guided" economies. The U.S., for the most part, is still trying to make the old models and infrastructure that made us a superpower work through cost cutting and more aggressive application. Western Europe's reaction has mostly been to form a trading block and tap Eastern Europe for cheap labor.

Though strategies vary, most agree that the 21st century demarcates a post-industrial or knowledge-based society, one where the fundamental sources of wealth will be knowledge and communication rather than natural resources and physical labor.[9] *Anything that is standard and routine quickly falls to third world wages and razor-thin margins.* In such an age, where capital and production capacity are globally abundant, endowed factors are increasingly unimportant.

In 1962, Princeton economist Fritz Machlup published the Production and Distribution of Knowledge in the United States. Machlup attempted to measure the economic value of knowledge production. Using 1958 data, he concluded that 34.5 percent of GNP in the United States could be allocated to the information sector.[10]

In 1977, Marc Porat, later CEO of General Magic, a software developer in Silicon Valley, wrote *The Information Economy* where he used a slightly broader aggregation and concluded that the information sector was 25.1 percent of GDP and generated 43 percent of all corporate profits. In 1992, IT and the communications sector had grown to 10 percent of the Gross National Product in the United States.[11]

The computer equipment and service sector alone was larger than automobiles, steel, mining, petrochemicals, and natural gas combined.[12] Today, a single company, Microsoft, has roughly the same market valuation as the total of the once preeminent Machine Age military-industrial complex.

The competencies underlying these three periods of history are illustrated by their various institutions. The scope of managerial competencies in the 17th to 19th centuries was focused at the national level. Much of the organizing mode was craft-like and operated under feudalistic institutions. This era of comparative advantage saw the early industrial revolution, though most trade was based on natural resources (e.g., tea, cotton, and opium).

As trade and commerce escalated during the 19th and 20th centuries, craft evolved into machine structures, with factories becoming more dispersed across national and regional boundaries. This was the machine era. The emergence of factories served the function of aggregating pools of labor under one roof, often supported by capitalists. Indeed, the factory with its belching smokestacks became the icon for Machine Age prosperity. Factor endowment based models were dominant, and one factor was technology.

The new era is the Information Age. In today's knowledge-based economies, there is much more attention to services — from software to innovation to art — as strategic components. The scope of operations is literally global, and moving to virtual or cyberspace. Factor endowment based models don't work well, if at all.

Where once industries could be classified in terms of distinct, corresponding technologies, this is no longer true. In the 1970s, single-technology products dominated the list: chemicals, petroleum, metals, plastics, and drugs. In 1990, complex multi-technology assembled products were prominent: consumer electronics, computers, cars, airplanes, and telecommunications.[13]

Today's products are not as they might appear to the uninformed. For half a century, the value-added to cars came from steel. No longer. Today the greatest value-added is in the area of software, electronics, and exotic materials. As important, this applies not just to the end product but also to the tools that build the product.

Industries and competition are now becoming defined in terms of dynamic, technological terms. The increasing pace of technological innovations has shortened life cycles and has made speed a crucial competitive weapon.[14] Network and architectural blocks have emerged as key competitive structures. In order to establish dominant designs, rival networks are forming networks. Competition is increasingly being fought between architectures rather than products.

Consider two popular examples of architecture battles: the almost religious rivalry between PCs and the Macintosh, and the freewheeling cyberspace melee of information and communication technology. In the personal computer battle, two traditional rivalries have collapsed into one. The first is the DOS versus Macintosh software-operating architecture. The second is the CISC computer chip architecture of the microcomputer market versus RISC architectures from the workstations and larger computers. Today it is "Wintel" (Windows plus Intel) against the world, and the battle seems all but over.

The highly touted information superhighway is an enfolding group marriage of the hardware configurations of digital industries (semiconductors, consumer, and electronics), the transmission industries (telephone and cable), and the software industries (publishing, video, games, and television marketers). This conflict is made more complex because of powerful outsiders and regulated markets.

There was no government involvement in the conflict between the PC and the Mac, but the Megastate closely regulates telecom and electronic media. This creates a new dynamic, one somewhat akin to pitting an irresistible force against an immovable object. Intel bemoans the fact that while the power of a PC doubles every 18 months, the bandwidth of the public telecom system doubles every 100 years.

Regulatory blockage is hard to change, despite all the hype, since powerful groups — broadcast and print media, Hollywood, and the local phone companies among others — feel threatened by truly low-cost high-bandwidth universal access to the Internet. Today, the U.S. has about 2–4% penetration of local fiber. Singapore, in contrast, has 100%.

Electronic commerce is impossible without free, private communications, but Megastates want control of cyberspace. The U.S. is doggedly insistent that its draconian export laws for nuclear weapons should apply to privacy software. Top Commerce Department officials have repeatedly attempted to make web browsing a possible felony, contending that the innocent act of viewing a computer screen violates copyright law because the computer is making "illegal" copies. Clipping articles to share with colleagues would be illegal, if done electronically.[15]

Though the press is staunchly upbeat, it is hardly inspiring that Washington has delegated responsibility for the Internet to Ira C. Magaziner, most noted for failed Megastate health-care and similar disasters. Digital television is experiencing levels of pork barrel politics and regulatory bungling that may prevent its deployment for years.[16]

If anything, Internet regulatory progress in the U.S. has been negative. In 1997 for example, the FCC awarded extra fees to the local phone companies for business and second residential phone lines. Therefore, the phone companies get more revenue from the Internet and still have no incentive to build data-friendly networks. There has also been constant conflict between regulators and citizens over the right to cyberspace free speech.

Still, in the end, the U.S. will somehow develop a 21st century infrastructure. In this competitive arena of cyberspace, the winners will not necessarily be the possessors of superior technologies, but rather those that can capitalize on their positions within strategic networks, and those who are able to master the ever complex patterns of renewal and growth that form the core study of this book. Basic features in the evolution of managerial competencies from factor and machine, to knowledge-based economies are presented in Table 1.

Table 2-1. The Transition to Knowledge-Based, Information Economy

Characteristics	*17th-19th Century*	*19th-20th Century*	*21st Century*
Basis of Competence	Factor-Based	Machine-Based	Knowledge-Based
Production Mode	Craft/Factories	Automation/Hierarchies	Services/Networks
Scope	Local/Regional	Regional/National	Global
Industry Classification	Distinct; single	Distinct; multiple	Diffused; architectures

Characteristics of the Information Age

What are some underlying characteristics of this new Information Age? In this section, we present key themes of this age as characterized by leading scholars.

Dematerialization

In his book, *The Intelligent Enterprise,* Dartmouth professor James Brian Quinn reports that information has become the source of about three-fourths of the value-added in manufacturing.[17] Advances in technology, particularly logistics, computer-aided design, and communications, have prompted companies to outsource factory work. Consequently, three out of ten large U.S. industrial companies outsource more than half of their manufacturing. Chrysler outsources 70 percent of its automotive products. Yet outsourcing is not limited to manufactured products alone. Money has dematerialized. Today, some $1.3 trillion in currency is traded every day, electronically and without ever taking tangible form.

Connectivity

In their book, *Paradigm Shift: The Promise of Information Technology,* Don Tapscott and Art Caston, leading scholars in the field, see major discontinuities between the Machine and Information Age.[18] Specifically, they argue that advances in information technology have driven the restructuring of national economies in major key shifts as follows:

Table 2-2. Continuities and Discontinuities

From:	*To:*
traditional semiconductor	microprocessors
host-based	network-based
vendor-propriety	vendor-neutral
separate data, text, voice, image	multimedia
account control	multi-vendor partnerships
craft	engineered
alphanumeric character set	graphical
stand-alone	integrated, systematized

Amo Penzias, the 1978 Nobel laureate in Physics, first theorized that the convergence of computing and communications had led to connectivity.[19] In effect, he theorized that everything will eventually be connected. In most of the examples provided by Tapscott and Caston, technologies have enhanced the connectivity of various elements. A high school student at Lane Community College in Eugene, Oregon could have as much access to information through the Internet as a Harvard MBA student. A stock trader in Montana with a satellite feed can have same timely data as his colleagues on Wall Street.

Connectivity facilitates transactions. Information *per se* becomes transactions.

Table 2-3. Shifting Themes

Theme 1: *Knowledge*	The new economy is a knowledge economy.
Theme 2: *Digitization*	The new economy is a digital economy.
Theme 3: *Virtualization*	Physical things will become more virtual, less bounded and specific.
Theme 4: *Molecularization*	Mass becomes modular, with more disaggregation and dynamism.
Theme 5: *Integration/Networking*	The new economy is a networked economy, integrating molecules into clusters that network with others to create wealth.
Theme 6: *Disintermediation*	Middle functions are being eliminated and need to create new value-added functions.
Theme 7: *Convergence*	The dominant economic sector is being created by others that provide infrastructure for the new wealth.
Theme 8: *Innovation*	The new economy is an innovation-based economy.
Theme 9: *Prosumption*	In the new economy, the gap between consumers and producers blur.
Theme 10: *Immediacy*	Immediacy becomes a key driver and variable in economic activity and business success.

Virtual Networks

In a sequel, *Digital Economy: Promise and Peril in the Age of Networked Intelligence,* Don Tapscott elaborates on the themes that undergird the new economy:[20]

The themes espouse a new organizational modality, that of a virtual network. Physical things become virtualized. As virtual corporations, teams, agencies, and units abound, then the metabolism of the economy, institutions, and the nature of the economic activity itself become transformed. Concurrently, the new economy becomes a networked one with deep, rich interconnections within and between organizations. The new enterprise becomes an internetworked enterprise, made possible by integrating modular, independent organizational components.

The Transition to Information Age Thinking

Business in the Machine Age was defended in depth by rigid, interlocking frameworks of tacit assumptions. Think of Machine Age managers and their firms as wounded organisms. They are in pain and desperately trying to survive. The problem is that the things that worked so well in the past just don't work anymore. Worse yet, even when managers somehow stumble upon the right things to do, their actions, guided by mindless process and past models, are often fruitless. Good intentions and hard work are thwarted by "the system."

Box 2-1

> **"Almost everyone agrees that the command and control corporate model will not carry us into the twenty first century. In a world of increasing interdependence and rapid change, it is no longer possible to figure it out from the top."**
>
> Dr. Peter M. Senge, MIT, 1997

Because the changes that characterize Information Age thinking can be subtle, managers mistakenly think that previous models, particularly when stretched sufficiently or applied aggressively, can still work well. That's wrong, though increasing control, reorganizing, and downsizing can often mask the symptoms of decline, at least for a time. Therefore, the first implication for those struggling to understand this new age is to be aware of *blind spots.*

Recognizing Blind Spots

Examples of blind spots are legion. They include IBM's preoccupation with mainframes, DEC's fixation on mini-computers, and Tektronix's failure to migrate with their loyal customers, first from analog to digital, and then from test-based design to simulation-based design. Such past-looking actions were appropriate, even optimum, for the markets of a decade or two ago, but wrong for the new age. Examples of correct, but fruitless, actions are even more common. They include IBM's successful technical development of mass market PCs and robust desktop 32 bit operating systems, which they (so far at least) failed to convert to market dominance.

DEC was the early leader in networked computers, but it allowed other firms to capture the markets for both networking and workstations. In the 1990s, their Alpha chips are technically superb workstation components, but that market has already committed to other solutions and is reluctant to change.

Tektronix was a high tech leader through the 70s. They had UNIX workstations based on mass-market chips in experimental use and limited production years before Sun's products. But Tektronix neglected the small, fuzzy, emerging, workstation market because there was a lack of "facts" to guide their Old Think decision processes. The delay cost them dearly, and they never really recovered.

Only after the workstation market had developed was the data that management wanted available. It was too late, but Tektronix spent lavishly trying to catch the leaders. Their efforts were expensive and disastrous. So it was with Tektronix's repeated forays into Computer Aided Engineering (CAE), and, to a lesser extent, into Automatic Test Equipment (ATE). Many hundreds of millions of dollars were wasted. GE, RCA, Wang, and many others had similar experiences as their markets changed.

Apple's Newton was early to market and technically clever, but misguided. Trying to push past the mismatch with aggressive promotion caused what *Business Week* called "a marketing and public-relations fiasco." Their attempts to encourage clones and port the Mac's OS to popular platforms were correct, but half-hearted and several product generations too late. The list is endless, but the root cause is usually the same: these firms are being killed by Old Think.[21]

Box 2-2

"Most companies our size that have been very successful got themselves in deep difficulty — IBM, DEC, General Motors, Sears Roebuck. The only mistake they made is, they did whatever it was that made them leaders a little too long. I worry that I too will be party to hanging on too long."

Lewis Platt, CEO, Hewlett Packard Corporation[22]

Diagnosing Barriers to Change

Machine Age organizations are incredibly robust. Their structure comes from 19th century military models. Most companies are machines where "the system," not empowered people with knowledge, dictates action. Structures, procedures, and lines of command are rigid. The system may be rusty and lame, but it rolls along inexorably.

Machine Age firms function long after savage downsizing and the elimination of experienced workers. The more they are damaged, the harder they struggle. Some large organizations have even demonstrated that they can still function after brain death, because they run mostly on routine. Jobs are narrow and prescribed in detail. Employees are told they are valuable, but they know they are the firm's most expendable assets. Ironically, knowledge workers, being more expensive, can be the most expendable of all.

What if the system itself is the problem? Decades of work to help traditional firms "manage change" have been dismal failures. Every large technology firm is ringed by spin-offs based on innovations it would not or could not pursue. The best people often abandon their careers at large firms rather than fight the system. Others "keep their heads down" and "play the game" to survive.

If there is one challenge for consultants, it is helping their clients make the necessary changes without excessive carnage and bloodletting. As Hamel and Prahalad have noted, the goal is a transformation process that is revolutionary in result, but evolutionary in execution.[23] The mantra for the Machine Age was "Obey the rules. Don't screw up, or else." The mantra for the Information Age might be stated as something like, "Innovate. Seize opportunity. Change the rules. Learn faster than your competitors." These two attitudes conflict, and successful Machine Age managers have great difficulty adapting to the new era.

It is amusing to study successful products, since those who get the credit are seldom those who started the work. Who remembers Jeff Rifkin? He was the early champion of Apple's Mac, fired when he complained to the board that Steve Jobs was a "dreadful manager."[24] Since new product failures often topple CEOs, it should not be surprising that some want to take personal credit for successes. Common wisdom attributes the biggest blunder in computer industry history to John Akers, who tried and failed to move IBM from mainframe computers to a dominance in client-servers and personal computers.

Conversely, executives or workers who follow routine and fail to seize opportunity often escape blame. Consider John Sculley, who, intimidated by Bill Gates, never even attempted to port Apples' superior graphical operating system to mainstream platforms and the Intel chip set. Sculley was eventually fired, but his dismissal is generally attributed to bungling the Newton or "bad numbers," not for missing earlier opportunities.

In the large Machine Age firms, career advancement depended more on not making embarrassing public blunders than on doing anything right. Severely downsized, dumbed-down firms often elevate their loyal process workers to lofty positions, regardless of their lack of ability. Leadership, innovation and creativity are valued less than obedience.

The Information Age calls for different behavior. The winners must experiment and take risks. Sins of omission (like Sculley's) are frequently more damaging than errors of commission. Minor mistakes in the heat of battle are inevitable, and those that lead to organizational learning are seen by leading innovators as merit badges. The Venture Capital community will often fund a CEO whose last venture failed, provided that he learned from the experience. Intel's Pentium bug cost $500 million, but the chip is still the most successful in the firm's history.

While Machine Age firms focused on predictability, success in the Information Age will necessarily focus on innovation and rapid learning. Learning at the cutting edge requires making quick intuitive decisions, taking risks, and, yes, making mistakes. Only through mistakes is rapid learning possible. In today's markets, early but imperfect products almost always beat late entrants, however perfect. This necessitates major changes in viewpoints and reward systems.

The Fallacy of Numbers

In retrospect, one of the more unfortunate trends in U.S. business during the 1980s was the shift away from an in-depth content knowledge of business to financial abstractions. Only by looking back, as in a time lapse photograph, is it obvious how much the philosophy of business has changed from content to technique.

In the 1950s and 1960s, successful companies were typically run by visionary builders. Consider David Sarnoff at RCA. He had a vision for a business centered on television. There was no market for color television receivers, because there was no such programming or broadcasting. In defiance of his entire board of directors, he built the needed infrastructure so that he could sell color TVs. It led to a prosperous company and a lucrative industry.

Sarnoff was perhaps an extreme case, but he was typical of the pattern. Hewlett and Packard did similar things. Starting with a vision for electronic signal generation, they created a highly successful firm with businesses that ranged from test and measurement equipment to computers. So it was with IBM in management information systems, with Collins Radio in wireless communications, and with Tektronix in the visual depiction of electronic phenomena.

In all these cases — and there were many others — the commonality was (1) identification of an unserved customer need, (2) exploitation of unique technology and know-how to meet that need, and (3) creation of a business to deliver products and services to a set of customers. This was the model for Western success until the late 70s, when it became the model for Japanese success. Sony exploited transistors to build small radios and Toyota exploited poor Western quality to build cars.

Box 2-3

In the Information Age, it is dangerous not to stay current with trends. Many think that the Japanese are defeating us because of quality. That is becoming wrong. It was once true, but the times and strategies are changing.

Nearly 80% of U.S. managers polled think quality will be a fundamental source of competitive advantage in the year 2000. Japanese managers disagree. While 82% think it is currently an important advantage, barely half think it still will be in 2000. 82% of Japanese manager's rank the capacity to create fundamentally new products and businesses as the primary competitive advantage by the year 2000.[25]

Significantly, the old models broke down as visionary founders aged and markets grew into ever more complex, fast changing niche structures. The founders with clear vision were largely replaced by technique-based management professionals. Mitch Kapor's user friendly spreadsheets (Lotus 123) provided easy tools, and Boston Consulting Group's (BCG) popularization of "portfolio management" of Strategic Business Units (SBUs) accelerated this trend. And so it went with the other sectors of portfolio management. As Sun Tzu said many centuries ago, "The best way to wage war is by attacking the enemy's strategy." Prepackaged strategies lead to predictable behavior that a clever adversary can easily exploit.[26]

U.S. strategic planners took the portfolio idea one step further. They developed formulas that appeared to identify the contribution each business element was making to a company's overall stock price. Called value-based planning, its application, along with junk-bond-driven leveraged buyouts created a feeding frenzy of acquisitions during the financial go-go years of the 1980s. (Yes, Japan suffered too. They blame us for devaluing the dollar by 250%, causing them major losses.)

Junk bonds and BCG portfolio management are largely behind us now, but their legacies persist. Since these planning methods are both logical and quantifiable, descriptive as well as prescriptive, they remain highly seductive to upper management. What an easy way for the head of a corporation to put his arms around what might otherwise seem a perplexingly complex and diverse array of businesses.

Peter Drucker did a short note in *Wall Street Journal* a few years ago saying that nothing on a financial statement measures the main things that govern today's business success. There is no column that measures market understanding or technology strength, and, in fact, deficiencies in these areas do not show up on a balance sheet until long after the damage is done. That seems obvious, but saying so violates dogma and even Drucker's brave comments died in silence.[27]

U.S. financial reports have become rigid prisons for executives: "How long have you been CEO? Five quarters." The top-level problem is not so much that U.S. financial reports are uninformative, it is that they mislead. Maturing and dying companies can generate a lot of cash and profit. Much as a person dying of AIDS may produce good x-rays, dying businesses can produce excellent numbers. (U.S. accounting systems were designed for paying taxes and counting industrial age assets. Attempts to overlay better financial metrics, such as activity based costing, conflict with these and with downsizing trends.)

Since most firms reward based on balance sheet numbers, is it any surprise that high CEO pay and corporate decline are increasingly common — and often linked? It is much easier to kill a business profitably than to run it well.

Box 2-4

> **"It is very difficult to discern (from looking at their financial reports) between well-run companies and profitably going out of business companies."**
>
> Arthur C. Lee
> Managing Consultant, EDS

A broader problem with financial abstractions is "confusing the map for the territory." While a map is a useful planning tool, it in no way helps one organize power and resources for a difficult battle. Just as tracking the gauges is a small part of piloting skills, finance is a small part of knowledge-based business. Many MBA programs now teach that the goal of business is to create profits. That view is totally backwards. *The purpose of business is to profitably create customers.*

The Japanese take a more business-oriented approach to accounting. Instead of corporate accounting departments, most firms have small groups of generalists attached to projects. They also flip-flop how they invest their resources. The U.S. typically assigns a fraction of a financial person when planning a new product, but four full time "bean counters" when it is done. Japan does the reverse. They will assign four senior people to help sort things out, and only a small fraction of a junior person to count things afterwards.

Box 2-5

> **"The Japanese don't think these financial statements are very important. You should take them in the spirit in which they are prepared."**
>
> Daniel Maher
> Partner, Chou/Coopers and Lybrand, Tokyo

Good Numbers and Bad Numbers

Much as pilots require their airspeed, altitude, and fuel status, managers need timely numeric information. Please don't think that Information Age management is only fuzzy, touchy-feeley, and soft. The leaders, in fact, tend to be both high tech and high touch. Think of modern financial trading. The industry uses state-of-the-art communications, software, and computer technology to monitor markets in real time. Firms invest in new technology lavishly, because the company that get the right answers a few minutes sooner does not just increase its profits — it makes *all* the profits. Such firms also invest in traders and analysts with good track records at using modern tools to intuit proper action. Winning takes both technology and talent.

Lester Thurow has said, "U.S. managers are under-scienced." That puts us at a major disadvantage to those who understand how technology can change products and business operations. New tools let experts make better use of their training and intuition. Human experts can use real time graphics — soon virtual reality — to better manage complex interactions in their businesses and markets.

Technology lets managers monitor operational performance in real time. Choosing the proper metrics and information systems can provide major competitive advantage. The numbers that matter the most measure internal processes and external reality, and they are not found on the financial report. In the end, these will determine how the business performs.

Converting operational data to timely knowledge gives competitive advantage. Naturally, all the knowledge in the world does not help if you still make the wrong decisions. In the Information Age, making the "right" decisions too late is wrong.

Recall Sculley's early blunder. Apple would have had major advantage had it licensed its operating system on the Intel chip set in 1988. Their advantage would have been similar to owning the only ice cubes in the desert. That is obvious in retrospect, but it's academic in 1998. The cubes have melted, so option no longer exists.

The knowledge to guide timely strategic decisions comes from deep understanding of the market and technology, not from finance. Compared to such decisions, tweaking the balance sheet matters little, at least not in the long run. The implications are major: traditional finance will almost certainly *not* be the center of the Information Age business solar system.

The difference between deep knowledge and financial abstraction is more than semantics. Generals who push symbols around maps usually lose to those who know the territory, their troops, the adversary, and the flow of battle at a detail level. Top management preoccupation with the traditional financial scorekeeping aspects of business can divert attention from the underlying mechanisms that create value. This can be fatal.

In the societies that lead the Information Age, the horse and cart will reach proper juxtaposition. The world of finance will again serve, rather than control, business. This is the Post-Capitalist society of which Drucker speaks.[28] The best structures to enable it are still evolving.[29]

Bill Gates cited a marvelous example of putting business prosperity before the quarterly report. Craig McCaw created the notion that the value of his company was linked, not to its balance sheet, but to its number of subscribers. McCaw Cellular lost money year after year on an accounting basis, but attracted huge investments. In the end, McCaw sold his firm to AT&T for billions.[30]

Putting business results before the financial report worked for McCaw. It worked for Gates, who invested lavishly in Windows. It worked for Andy Grove at Intel, who invested in a treadmill of ever improved replacement products and new "fabs." It also works for the leaders in prosaic industries.

Wal-Mart has produced stellar results in the most mundane of industries — mass merchandising. Industry margins are depressingly low, and competition is ferocious. Wal Mart did not win because it manipulated its balance sheet better. They won because they exploited general aviation, multimedia communications, and computerized inventory control and fleet management to become one of the fastest growing and most profitable companies in the world.

Emerson Electric is performing well in industrial components, traditionally a low margin backwater. It delivers excellent business results because it is focused on not just operations and cost containment, but also on new product innovation. It too uses Information Age technology aggressively, and is now a thriving $11 billion global company.

Smaller firms can exploit knowledge-based advantage as well, or better. A $200 million Portland, Oregon firm called AMERICOLD has come to dominate frozen food shipping by applying proprietary software to optimize shipping transaction costs in near real time. Many of its former competitors have become customers.

The Return of Seasoned Intuition

New age leaders can be trained in how to make intuitive decisions and what type of mistakes are permissible. Some mistakes — the "below the waterline type" that could sink the company — are forbidden, but "failing quick and cheap" must be allowed to facilitate rapid organizational learning.[31]

Information Age firms cannot prevail without allowing mistakes that permit rapid learning. Engineers know this intuitively — those seasoned in new product development often push early experiments past the point of failure for learning purposes.[32]

Still, intuition depends on timely and correct knowledge. A common complaint from Information Age managers is that they are, "Drowning in data, but starved for information." Knowledge is even more elusive than information. The most crucial decisions often must be made intuitively, and before all the facts are known. Intuition can be taught, as can real-world risk management, though the model of the Guru or Zen master is better than that of the classroom.

Conventional education is a large part of the problem. Traditional MBA training forces managers into frantic data collection. It focuses them on financial abstractions, and it moves them at blinding speed through terse summaries and rigid analysis to urgent lists of action items. Such managers live in a world of sound bites and crisis, and they often have grave difficulty in perceiving the world (indeed, their lives) differently.

Box 2-6

"If calculus was invented today, our organizations would not be able to learn it. We'd send everyone off to a three-day intensive program. We'd then tell everyone to try to apply what they had learned . After three to six months, we'd assess whether or not it was working. We'd undoubtedly then conclude that this 'calculus stuff' wasn't all it was made out to be and go off and look for something else..."

Quote from a Ford manager at MIT[33]

A principal assumption is that problems are "convergent," that honorable and competent people who apply proper analysis to "the facts" will always reach the same answer. This is nonsense. Many important problems are *divergent*, and good people will reach opposite (and correct) conclusions about what to do — how to best educate children is a good example.

The approach to "effective communications" is called single-loop learning. MBAs are trained to communicate by bullet items and executive summaries. This works for some types of problems, but not for others. The approach blocks deep knowledge, and alternatives exist.[34]

Box 2-7

> **"If Edison had been an MBA, he would have invented a large candle."**
>
> John D. Trudel[35] *High Tech with Low Risk*

Under the old model, the workers should collect and package information for those blessed with advance business training, and then obey orders. Experience is irrelevant. Frederick Taylor, the father of Machine Age management liked to say, "The workers need be no smarter than an ox." Five decades later, his ox has been supercharged, whipped, downsized and equipped with blinders, but business methods remain the same.

A greater problem is that business training focuses is on traditional financial metrics — the bottom line — which confuses cause and effect.[36] In the end, does not the rich detail of intimate technical and market knowledge create the bottom line?

In many ways this is a fast food approach to gaining knowledge. It may be filling and it may be cost effective, but it is neither particularly tasty nor healthy. This approach frees workers from the need for thought, saving time and training costs. It promises increased predictability, especially if the minutia are all monitored and controlled. It gives the appearance of safety and of knowing the answers.

Box 2-8

> **"School trains us to never admit that we do not know the answer, and most corporations reinforce that lesson by rewarding the people who excel in advocating their views, not inquiring into complex issues."**
>
> Chris Argyris, Harvard Professor

Observing NASA's decline from a committed band of innovators into a bloated collection of fearful bureaucrats is an extreme example of the high price being paid for Machine Age management. Thus was the space shuttle launched with a defective gasket. Thus was the Hubble telescope launched with everyone blameless, but a flawed main mirror. It is unlikely that NASA could repeat the moon mission of 1969 today.

Box 2-9

> **"We'd have to borrow the money from Japan, and Clinton couldn't decide where to land."**
>
> Thoughts of a typical U.S. citizen.
> Jack McLarty, 27, from Chicago. Interviewed at the Smithsonian during the 25th anniversary of the moon landing.

There are other barriers to learning. Most Machine Age firms are heavily defended fortresses. Only insiders are privy to information. Rigid chains of command keep people from talking to anyone outside their department. Protocols define who can be consulted, advised, or criticized.

Box 2-10

> **"We are afraid of what would happen if we let... elements of the organization recombine, reconfigure, or speak truthfully to one another. We are afraid that things would fall apart."**
>
> Margaret J. Wheatley. *Leadership and the New Science*

Ms. Wheatley attributes managers' frenzy for order to seventeenth century science that viewed the world as a precision machine set in motion by God, who left us responsible to keep it running. This is a fearful legacy, the fate of Atlas with the world on his shoulders.[37] Stephen Covey also comments on the strangeness of corporations that, long after the mission is irrelevant and the reasons are forgotten, are still feverishly hacking though the same jungles that were discovered so long ago by their founders.[38]

Box 2-11

> **"One of your underlying themes is the message that 'You can't go back — or you die. You can't stand still — or you die. If you go forward, you may die, but you also have a chance to survive, and maybe even prosper and have some fun.' It is a good message. The trick is for a manager to do this in an old-line, traditional, industrial age firm."**
>
> Bob Draeger, Manager, Brown and Root

The Dynamics of Learning and Unlearning

The transition into the Information Age is still painful and embryonic in the West, but it will happen. The core concepts of Information Age management can be traced back to Mary Parker Follett in the 1920s. She died ignored. Her work was buried and not rediscovered until recently.[39]

W. Edwards Deming, the person who helped Japan though this transition — but failed to convince his native land — suggested leadership over control decades ago. Unfortunately, our adversarial Machine Age culture ignored him and distorted his teachings. He is portrayed, incorrectly, as a quality process worker, though his writings pleaded for leadership and deep understanding. In his lectures, he endlessly repeated his mantra: *"Leadership means removing the barriers that prevent people from taking pride in their work."* He is associated with TQM, a process solution that he never advocated and a term he never used.

Box 2-12

> **"A pattern emphasized ... is the degree to which powerful competitors not only resist innovative threats, but also resist all efforts to understand them."**
>
> James Utterback
> *Mastering the Dynamics of Innovation*

One key understanding in the Information Age is that the creation of technology does *not* drive wealth. What drives wealth is the proper *application* of technology. The Japanese know this well, but the U.S. is still learning.[40] Better understanding would make managers crave better marketing and more holistic viewpoints.

We need a new skill set — call them *Gurus* — who can blend technology knowledge and market knowledge at the cutting edge of change.[41] These are rare, but such people do exist. Generally they are engineers and scientists who have been cross-trained in marketing and business. Some are famous, but probably not available to you (e.g., Bill Gates, Akio Morita, Andy Grove, etc.).

Education is again a problem. Most of what is taught in the marketing curriculum comes from the industries that invest the most in marketing and promotion of consumer products. Compared to high tech, much more of what makes or breaks a consumer product is raw positioning and promotion. It is worth major promotional investments to shift a few points of market share for a soft drink, and you can do that with promotion and positioning, usually without touching the product itself.

Box 2-13

> **"The ability to learn faster than your competitors may be the only sustainable competitive advantage."**
>
> Arie De Geus,
> Head of Planning Royal Dutch Shell

Information Age business is an art, not a repeatable bench science, and it can be fun. There are no rules, and the technology and markets are inexorably intertwined. The process of creating a winning product is highly content dependent. It takes intimate, subtle, high content, interactions between team members, and extending to customers and suppliers. For example, just knowing who to interview to define a technology-based product is a fine art. What to say, or ask, is harder. Understanding and effectively applying the information you collect — converting it to knowledge and informed action — is so difficult that few firms do it consistently well.

A Silicon Valley new venture called GO corporation learned a hard lesson, one now cast in high tech folklore as the Valley's "date rape" case. GO had a vision for a new type of computer that used a pen, rather than a keyboard, as an input device. They visited Microsoft for market research under a nondisclosure agreement, talking through the technology and application tradeoffs in a friendly manner.

The result? Microsoft did not *disclose* anything, but they did develop a pen based computer interface of their own. Instead of a customer or a strategic partner, GO developed a ferocious, established, multi-billion dollar competitor who now knew their technology and strategy intimately.

This was clearly a major marketing blunder on GO's part, but other firms in other industries have made worse mistakes. It was not their only blunder. GO is gone, and the market they imagined has yet to emerge.

In the Information Age, the most valuable property is intellectual property. Should it be any surprise that the theft of intellectual property is escalating at a high rate, up 323% from 1992 to 1995? FBI director Louis J. Freeth testified to Congress in 1996 that most estimates place the U.S. losses "at billions of dollars per year." Illegal knock-offs of products are increasing as fast. *Business Week* said in 1997 that losses were $400 million per year, a 400% increase in the last decade.

Since one makes money by applying, not by creating, technology, precautions are in order. A San Diego new venture, Celeritas Technologies Ltd., experienced the same type of "date rape" as GO, but they were prepared. In this case, the technology was advanced communications modems. The suitor was mighty Rockwell International, who also was courted under non-disclosure agreement.

Rockwell declined to license the technology, apparently choosing to simply copy it. But in this case, Celeritas had patent protection and sued. In 1997, the courts ruled against Rockwell. Celeritas was awarded triple damages plus their legal fees. The jury awarded over $56 million, and the judge, who said some witnesses "looked like they were lying," tried to double it. This signal event is causing all the players in the modem market to pay closer attention to intellectual property ownership.

It works the other way too. A large Silicon Valley firm, Cadence Systems, is suing a smaller competitor, Avant, for theft of trade secrets. If convicted, Gerald Hsu, the CEO of Avant and a former Cadence employee, could face seven years in jail.

Even large firms, who formerly utilized their patent positions like the U.S. and Russians used their Cold War nuclear warheads (for purposes of intimidation, not outright war), are flailing away at each other in the courts. At this writing DEC and Intel are fiercely contesting who infringed whose microprocessor patents. (The only thing that all parties seem to agree to — except, of course, the lawyers — is that creaky old Western property law works better against cattle rustlers than knowledge thieves.)

Intellectual property theft now involves governments as well. China has long been accused of turning a blind eye to software piracy. Japan in 1994 used trade channels to secure letter agreements with the U.S. Commerce Department which, if approved by Congress, literally pirate the U.S. patent system, allowing cheap access to U.S. technology.[42]

In any case, business in the Information Age is deadly serious trade war with billions of dollars and thousands of jobs at stake, not congenial academic discussions about potentially interesting applications for new technology. This is hardball, and the book about Bill Gates, *Hard Drive*, should be "must reading" for would-be Gurus.[43]

Innovation in the Information Age

Of course, having technology is of little value without knowing where and how to apply it. There is persuasive evidence that economically successful innovation comes from market knowledge. To quote Utterback of MIT, "From 60 to 80% of important innovations in a large number of fields have been in response to market demands and needs. Winning most often comes from 'outsight,' not insight."[44]

Technology firms have gone thorough cycles as they have groped with how to meet their marketing needs. Before the mid-70s, there wasn't much marketing at all. You just did good technology and it "sold itself." Since technology was scarce, selling was easy.

In the late-70s, you found formal marketing groups, but usually they were staffed with failed engineers. Customer support was a career path for manufacturing technicians. It didn't matter, because the engineers still did the products.

In the mid-80s, firms got more sophisticated. They now had all the trappings of marketing, spent lavishly, and touted their newfound awareness of customers. Many used MBAs, but they often tended to become preoccupied with numbers and rarely bothered to develop any deep understanding of customers, applications, and technology.

The engineers liked that. They still did the products, but something was starting to go wrong.

All too often customers refused to buy. New product success rates were falling. If marketing had the power to cancel or veto projects, it usually lacked the knowledge to guide them. This escalated the finger pointing wars that are so common in Machine Age functional organizations.

"We used to succeed before we had all those bozos in marketing. Why don't they just sell the bloody products?" snarled the engineers. It was a tough time for marketing managers, who found selling to be a challenge when the dogs would not eat the dogfood.

A newer model is that high tech Public Relations (PR) firms have managed to position themselves as the major marketing resource. Positioning substitutes for strategy, planning, and innovation. The ploy has been successful, at least for the PR firms, and sometimes for their clients.

This started in about 1984, and the noted success was Apple's Mac. Apple's PR agency, Regis McKenna, Inc., adapted consumer product positioning and distribution techniques to high technology business. It positioned itself as a marketing firm, and took most of the credit for the Mac's success.[45] The CEO of Regis — not too humbly, since his name was the same — wrote a book with a golden cover called *The Regis Touch.* The book's subtitle was "million dollar advice from America's top marketing consultant."

Enough customers bought the Mac to give 10% market share. Share was tenable, the product was cute, and extreme brand loyalty resulted. This was at a time when most computer industry products, including the previous Apple III and the Lisa, were failures.

Positioning-centric marketing still works sometimes. In the land of the blind, the one-eyed man can be King.

If you invest late in learning the market, you are stuck with what you have and good positioning is about the best you can do. Since most firms do invest late, the method is popular. At least you do not have to fire your marketing managers as often, and the engineers still get to do the products.

That this method frequently delivers limited results — fails — is testified to by *Crossing the Chasm,* written by former Regis partner Geoffrey Moore. Even if well positioned, the available market for techno-centric products may be quite small.

Recall the Newton example, where aggressive positioning and promotion by experts made things worse. Much greater losses resulted from the PowerPC debacle. The best PR firms and Ad agencies on the planet helped with positioning. The product bore the imprimatur of IBM and Apple and Motorola. Unfortunately and *predictably,* it sunk without ever putting the slightest dent in the PC market.[46]

Despite massive investments and good technology, the Power PC was all but forgotten in a few months. The author's prediction of this outcome in a national column was dismissed via a letter to the editor from Regis McKenna, Inc. The sad thing was not denial or wasted billions, but the missed learning opportunity. The key learning is that repetition does not work for knowledge-based competition, *because results depend on context*. The methods that worked so well for the Mac failed for Newton and PowerPC.

The other useful lesson is that positioning is only one component of marketing, and marketing is only one component of new product success. Just as all the positioning in the world would never help phonographs compete with CDs, it could not help DEC's Alpha chips or IBM's OS/2 compete with Wintel. "Promoting a bad product well will cause it to fail sooner."[47]

Winning today at new products takes more than positioning. In fact, it takes more than doing any one thing well.

The conclusion *not* to draw from all this fumbling is that new product success is impossible. Apple's next CEO, Gil Amelio, did little or nothing in the new product area, focusing, as do many, on deals and downsizing. He didn't last long, and he left Apple in desperate shape. Success is possible, as many firms demonstrate daily. It just takes different methods and new understandings.

Redefining the Customer

Traditional products have never worked very well for the emerging, dynamic markets that are so crucial today. Some of the biggest failures we have seen were associated with elegant market segmentations and reams of data to "prove" how good the product fit is to customer needs. Conversely, some of the best products we know of — e.g., Sony's Walkman, Apple's Mac, Intel's 386 — could never have been justified by conjoint analysis or by any type of quantitative, statistical market analysis.

Customers can't know that they need something until they know it exists, understand it, and have experienced and tested its applicability to their needs. Can you imagine Fred Smith, who founded the highly successful Federal Express company, doing questionnaire research for his new firm?[48] ("Sir, would you pay $10 to deliver overnight what the post office does in two or three days for a dime?"). Smith conceived the company as a term paper for a marketing class, and was lucky to get a "C." He laughed all the way to the bank and is now a multi-millionaire.

In 1900, Mercedes Benz did a market research study, proving that "the worldwide demand for cars would never exceed one million units, primarily because of the limitations of available chauffeurs." By 1908,

Henry Ford had introduced the mass market Model T, and by 1920, the U.S. alone had over 8 million cars on the road.

Box 2-14

> **"No customer ever asked for electric lights or photography."**
>
> W. Edwards Deming

High technology marketing has always been intuitive, and most successful professional practitioners know this. Technology-based products are very subtle things. The most useful knowledge is gleaned from analyzing high content discussions with the "right" people. Tools can help, but talking with a few visionary experts can be better than collecting statistics. In the end you seek knowledge, not data.

If a person buys a personal computer, they typically spend much time reading reviews, talking to experts, and researching technology trends. Some of the best market segments are power users who want the "hottest boxes," but will FAX twenty sources and take the lowest quotation. They almost never visit a computer store, because they know more than anyone they are likely to encounter there. They tend to ignore or discount advertising claims.

No one would go to that much trouble when buying a soft drink, and few do it when buying cars, but such a selection process is common in high tech. If the purchase involved is a LAN for a company, or a fleet of aircraft, or a supercomputer, the selection criteria would become even more rigorous.

What matters in such markets is content knowledge, not statistical data. One smart, informed person who understands how technology and competitive forces will lead the market outweighs a thousand uninformed or misguided survey points.

Stuart Alsop, the computer industry columnist once said, "The only way to make a lot of money is understanding a customer need before the customers are aware of it, and then bending technology to meet that need in a way that is uniquely yours." He is correct.

Casey Powell, the founder and CEO of Sequent Computers, uses a football analogy we like. "To play in the major leagues you have to throw the ball before you see the receiver." This has several meanings, all of them helpful. For one thing, the quarterback cannot stand back and "manage" the team in the typical Machine Age business fashion. If he does, his team will lose. Instead he has to do his own job well, and he also has to trust his well-trained team to perform instinctively.

We in the U.S. mistake "trained" as synonymous with "exposed to knowledge." For example, "I took a class, read a book, and got a certificate. Now I am trained in digital circuit design." That type of knowledge is trivial and not what we are talking about. Our definition of training goes beyond this and requires the actual repetitive practice of knowledge. One becomes a concert pianist or a professional sports player only through a combination of learning, innate talent, and constant practice against increasingly more difficult competitors.

Box 2-15

"The key to developing excellence is active participation."

Stephen Jay Gould, Harvard University.

This poses a choice for managers, one that needs to be revisited. In a knowledge based economy, should seasoned talent with good track records be sought out, cherished, and retained for future advantage? Or should it, being expensive, be shed at the end of the project or the next layoff, to improve the balance sheet?

In sports, of course, managers choose the former. The quarterback has a difficult task. Adapting in real time and after misleading the defense, he throws the ball, hard and accurately. Then, if everyone on the team makes their blocks and runs their patterns perfectly, there will be a receiver in the right place to catch the ball **and** he will have a three-step lead on his opponent. Success depends on skill, team behavior, trust, and risk taking. So it is, we submit, with business in today's Information Age markets.

In computers and other Information Age industries, one must often start product development before anyone *knows* and can prove the market need. More important, however, is knowing that every one of your opponents is smart and well trained. Your competitors all read the same surveys, attend the same conferences, and have good technology. They are as smart and as quick as you.

The winners move beyond quantitative data to qualitative data, seeking competitive advantage. The process is enlightened discussions between the right people. Eventually everyone will target new business opportunities, but the winners are those who can consistently do it early and well.

One of the authors well remembers attending the prototype event for a leading edge executive workshop. It was a "skunk camp" that featured famous speakers, and a lovely location. The students were talented, and the group was mixed equally with executives from leading firms in both traditional business and high technology. The event occurred during the halcyon years of the early 1980s. It was a time when the U.S. still led the world in knowledge-based products. The mass layoffs were still ahead of us, but the long slippery slide of U.S. business was already detectable.

Tom Peters' standard exhortations for the audience to "love their customers" prompted a plaintive cry from the Vice-President of a large and already troubled computer firm. "But that advice doesn't help! It's not that easy. We are at the limits of our understanding in technology, and our customers are at the limits of their understanding in how they might use our products in the future. What should we do to gain the knowledge we need?"

Tom didn't like the question at all. He attacked the questioner, the unbeliever, with passion, vigor, and the utmost intensity. Some from slower change industries also had trouble relating to the question, perhaps thinking this sincere plea for help was only an excuse for poor performance. Those from high technology backgrounds were mostly sympathetic. The management of the high tech firm that had co-sponsored the event, bringing their best clients, one of whom was now under personal attack, was horrified. The subsequent interactions were amusing.

Still, reflecting back later on the experience was sobering. *The best minds in the country did not know how to stop the rain. There was no new magic*. Even if your customers were loved, that was not enough — because they didn't have the answers either. In the end, we decided we would have to get wet and work through events ourselves.[49]

Box 2-16

> **"Customers are notoriously lacking in foresight. Ten or fifteen years ago, how many of us were asking for cellular telephones, fax machines and copiers at home, 24-hour discount brokerage houses, multivalve automobile engines, video dial tone, compact disk players...?"**
>
> Gary Hamel and C.K. Prahalad , *Competing for the Future,* 1994

Living in the Information Age

One lives with chaos every day if competing in Information Age markets. No one really *knows* for certain which incarnations, of say, multimedia (intrinsically a mass-market consumer product), will prevail, but those who intuit correctly stand to make billions.

Some Venture Capitalist humorists now refer to Internet start-ups as *flies*. "You give them some money, they buzz around for a while, and then ***splat!*** The market swats them into a greasy little stain." This is black humor to be sure, but it also serves to educate.

Many of those who are funding Internet or multimedia firms admit that they don't really expect their first investments to pay off. They just want to get close enough to the action to know where to invest when sizable markets start to emerge. The experts are unsure exactly what or who will win, but they do know two things. Someone will make a lot of money. If they don't place their bets and play the game, it won't be them.

Such chaos and such "knowledge gap" opportunities have now spread to prosaic businesses like hotels, fast food chains, mass merchandising, and textiles. One of the most successful businesses we have seen, Hospitality Franchise Systems, is in the traditionally low margin hotel business. It owns no properties and maintains no staff to clean rooms or make beds. Instead it runs sophisticated Information Age reservation systems and conducts the training for most of the major hotel franchises in the U.S.

Box 2-17

> **"There is no objective reality out there waiting to reveal its secrets. There are no recipes or formulae, no check lists or advice that describe 'reality.' There is only what we create thorough our engagement with others and with events."**
>
> Margaret J. Wheatley *Leadership and the New Science*

Intel calls the data collected during early customer requirements interviews "murmurs." We like the term. It implies good minds focusing intensely and for long periods to fully appreciate barely discernible information in a sea of noise. In such a morass of data there are valuable nuggets of information, but it takes patient listening, time, investment, and great expertise to convert the data to knowledge for competitive advantage.

Gaining useful understanding in chaotic, complex, knowledge intensive, fast changing, high uncertainty, Information Age markets is very difficult. The new science shows us that there are patterns in chaos, but the best minds are just now learning how to deal with this.[50]

While Newtonian Physics had objective reality, the new quantum physics is an art, a dance, where nothing is wholly knowable and the observer and the design of the experiment shape the results.[51] The dance and the dancers are so intimately interwoven as to be indistinguishable. So it is with Information Age business. However business may leverage technology, it is subtle art.

Box 2-18

> **"Knowledge constantly makes itself obsolete, with the result that today's advanced knowledge is tomorrow's ignorance."**
>
> Dr. Peter F. Drucker

It takes more than desire to find breakthrough opportunity. It takes investment, skill, training, and a lot of hard work. Often it takes luck, and — as in baseball — even the best practitioners strike out frequently. The trick is gaining understanding *before* the key events have transpired and the answers are obvious.

As a final comment, getting the right products is not science but difficult art. Even with Gurus, it is a team exercise. Many firms have apparently abandoned hope of *ever* getting the targeting right for new products. Recall that marketing books like *Crossing the Chasm*, written by seasoned practitioners, pretty much accept that first products will be designed by and for technology enthusiasts, and will, hence, *always* mistarget mainstream markets.[52]

Still, even though today's best markets are often difficult, moving targets, it is not necessary to beat your brains out trying to position misfit products. Neither is it necessary to spray machine gun fire in all directions, or use the "ready, fire, aim" methodology. (Cute picture. You fire, paint a bulls-eye around the hole, and congratulate yourself.)

It is possible to get consistent results, and it is even possible to build "innovation engines" that help teams work together effectively. (Chapter 8.) The "simple" fact is the firms that can get the right talent to work together to get products done "right enough" the first time have enormous advantage.

Box 2-19

> **"Excellence is not an act but a habit."**
>
> Aristotle

In today's markets, such firms should win most of time. Better yet, their "failures" tend to be inexpensive learning experiences. Best of all, such teams tend to improve with experience, so you have still more advantage when you do your second or third product.

Chapter 3

The Dawning of Global Markets

Key Themes at a Glance

The second of our three challenges facing 21st century managers is the globalization of industries and markets. Globalization refers to the *deepening linkages and interconnectedness of economic activities on a worldwide scale.* Key drivers, such as the growing homogeneity of demand and the availability of supply factors are transforming semiconductors, televisions, financial services, machine tools, robotics, software and others into global products.

To effectively compete in a global environment, one has to develop a global mindset. In many ways, the United States has fallen victim to its own prosperity. As foreign firms played "catch-up," U.S. firms set the standard for domestic and international markets.

This mentality has produced unexpected consequences. The label, "Made in America," led to a disparagement of foreign goods. No longer justified, this preoccupation with American goods has led to biases against foreign, especially Japanese, products.

Oftentimes, our reception to superior foreign goods is articulated in charges of foul play and protectionist politics. Our scorecard on our educational system reveals deep gaps in bridging our understanding of foreign cultures.

This conflict is manifested strongly in relations with Japan. Trade surpluses often brings out feverish protectionist tendencies on both sides. In this chapter, we describe how Japanese have

developed their competitive advantages, through business — government relations, interfirm collaboration, and management and educational systems.

As with the Japanese, other Asian countries view institutions as a mantle for international competition. Lower labor costs, Confucian values, and "connections" find their way into intricate business relationships.

It is said that Adam Smith's "invisible hand" has given way to the Asian "hand-shake." It is evident that we need to learn much more about "cooperation as the future form of competition."

The pronounced significance of the "Overseas Chinese," globalization, and technological diffusion are other trends that are continuing to define the Asian competitive landscape. It is essential that these trends be understood, if not appreciated.

The economic and technological foundations of the postwar era are in fundamental transition. Competition in the world economy and market/political liberalization require new strategies and policies to promote growth and equity. The Asia-Pacific area is emerging as the world's fastest and most economically dynamic region. Growth rates of Southeast Asian economies are among the highest in the world. Many of these economies are compared favorably with the post-war economic performances of Japan, South Korea, Taiwan, Hong Kong, and Singapore. Greater China now represents one of the fastest growing countries and is the tenth largest exporter. That a financial bubble had led to an Asian financial crisis and a reassessment of growth strategies should not nullify the success of these economies in the past, nor should it detract from the significance of these countries as future global competitors.

Even so, its success has also raised concerns about its integration into the international economic order for the rest of the world. A handful of U.S. firms still enjoy a strategic advantage in some industries, but most have retreated in the face of escalating Trans-Pacific competition. Some traditional U.S. markets, e.g., commercial aviation and machine tools, are being challenged by European competitors who see their resurgence based on integration initiatives and a common monetary currency.

When the new industrial transformation is completed, the world will be very different. There will be massive reallocations of wealth and power. Some experts predict that the more traditional demarcations of industry and national boundaries will disappear. All over the world, national prerogatives are being ceded for trade advantage. Leaders are trying to adapt to chaotic new realities. There is heightened interest in business and policy circles in exploring how economic development in the region can be broadened, deepened, and institutionalized.

Not surprisingly, international competitiveness is becoming the most important economic issue in the world today, although there exists major disagreement on how to define it. There is even more disagreement on how to measure it.[1] Many say the West will be more changed by the end of the cold war than Eastern Europe. Our current assumptions and methods of competing will be fundamentally altered. Western management methods and organizational structures will be transformed in radical ways. Visionaries expect that by the end of the decade national competitiveness will be the major political issue in industrialized countries.

These economic and technological changes will usher in a new era — one characterized by "borderless" industry boundaries, cooperative forms of competition, information-based competitive advantage, transnational corporations and inter-industry networks, and new strategic trade theories.[2] An emerging reality is a new type of competition that is not based only from traditional rivals, but from the disintegration of national borders and previously protected national borders. As a consequence, competition from substitute products has become compelling. Moreover, new competitors, with different methods and ideologies regarding rivalry, have entered into the fray. Technological advances have all but equalized key international rivals, reducing the lead time for new innovative products and services that had once led to competitive advantages. The restructuring of national economies is relentless. Globalization presents a significant challenge to the Old Think Mindset. We discuss these forces driving globalization, their impact on how U.S. firms conduct business, and the challenges they bear for the citizenry in general.

Box 3-1

> **"The empires of the future are empires of the mind."**
>
> Winston Churchill

The Coming of a Global Marketplace: A Historical Perspective

The economic history of the world following the Second World War resembles a roller coaster.[3] The years following 1945 were ones of basic reconstruction of the war-damaged economies throughout the world. In fact, continual growth in world trade and production occurred from the 1950s to the 1970s. A particular feature was that trade increased more rapidly than production, signaling the growing internationalization of economic activities. Economies of the free world flourished and the standard of living improved in most developed countries.

This period growth came to an abrupt halt in the 1970s — a decade considered as "bust" years. While the popular view attributes this to OPEC's decision to raise oil prices in 1973, there had been a number of changes that were taking place throughout the 1960s that contributed to this demise. Commodity prices had been rising steeply. Labor costs in most industrialized countries also spiraled. International currencies became increasingly unstable as more and more national currencies moved out of line with fixed exchange rates.

As the recession deepened during the later part of the 1970s and continued on to the 1980s, it became apparent that a fundamental shift had, in fact, occurred. Corporate competitive advantages shifted to those of low cost, superior quality, and premium quality. With the tightening of world markets, attention focused on the blight of industrialization and the promise of Japan and the newly industrializing countries. Such attention was also followed with controversy as free-market proponents wondered how Japanese firms could have achieved competitiveness in so short a period. Before long, criticisms of so-called "industrial policy", or generally, governmental interventions and support of private enterprises had surfaced.

The emerging political and economic order of the 1990s (expected up to the year 2000) is slowly taking shape. Competition has escalated rapidly and pervasively around the world. Technological advances and the globalization of products has undermined traditional definitions of industry boundaries and the geopolitical affiliations of corporations. Economic blocs are beginning to take place. These hold promise for member nations, but the jury is still out on how these will impact inter-bloc trade. The era of growth between 1945 to 1985 is now being replaced by volatility and fast-paced change. Some refer to these conditions as the era of *hypercompetition.*[4]

Characteristics of the New Global Marketplace

Globalization is a word in good currency. Even so, it means different things to different people. In the 1970s, it was fashionable to use the term "international" to distinguish between domestic and foreign economic activities. Therefore, it described activities beyond one's national borders. Some firms still use "international" and "global" interchangeably, but the distinctions are both subtle and significant.

Globalization refers to the *deepening linkages and interconnectedness of economic activities on a worldwide scale.*[5] When one speaks of the globalization of semiconductors, for example, it invariably means the production and marketing of the product all over the world. It means that the product is uniformly accepted in parts of the world it is used. It means that it is globally sourced and produced worldwide. In contrast, other products that are still considered to be multi-domestic are limited in one or more of these characteristics. Telecommunications is considered by governments as a correlate of national security, and these governments have tended to rely on local suppliers. As a consequence, there is a wide variety of standards in telecommunications today: Spain has a three-second busy tone, while Denmark has two-seconds. France's telephones have seven digits, but Italy's can have any number.[6] Telecom is multi-domestic, not global.

Another way in which to view globalization is a platform in which players from major parts of the world compete or cooperate. Unquestionably, semiconductors represent a global industry where the Japanese, Americans, Europeans, and Koreans have emerged as key competitors. Lurking in the background are countries such as China, Thailand, or Taiwan that can enter the competitive fray at some time in the future. As such, there is an escalation of competitive or cooperative activities that are partly manifested in intra-industry trade ratios and foreign direct investment.[7]

What sets the pace of globalization?[8] This is determined by underlying industry conditions and characteristics. To achieve the benefits of globalization, managers of worldwide corporations have to recognize when such conditions provide the opportunity to develop a coherent strategy. We define these conditions and characteristics accordingly.[9]

A Trend Toward Homogeneous Worldwide Demand

Homogeneous customer demand represent the extent to which customers in different countries develop similar needs in the products and services that define the industry. In many areas of food, fashion, and entertainment, the world is growing similar tastes and requirements. The demand for McDonald's hamburgers, Coca Cola, Benetton clothes, athletic shoes, consumer electronics, and homogeneous technical standards are creating "global" products.

The increase of consumer incomes throughout the developing world, coupled with increased accessibility to these trends through telecommunications and travel, has facilitated greater consumer similarity in terms of their taste and penchant for global products. Japanese electronics, records and discs are becoming more standardized, even with some minimum amount of local adaptation (electrical voltage plugs). Medicine is among the most homogeneous of products, spawned in part by rapid transmission of information via the medical journals. Even food products, considered to be national in origin, are becoming widespread. Japanese have been converted to eating doughnuts, British are more accustomed to drinking cold American and European-style lager, and Americans have developed a hearty appetite for *sushi* and French mineral water.

Lower Global Transportation Costs

A major decision confronting corporations seeking to enter international markets is whether to export or to build foreign subsidiaries. In this context, transportation costs of raw materials, components, and finished goods play a key role. The heavy weight and large size of some products result in transportation costs that would discourage shipments. But, in other products with low weight and high value-added, such as consumer electronics, the economics of centralized manufacturing and low transportation costs make it preferable to manufacture centrally.

Global transportation of many commodities costs have fallen since the 1960s. This has facilitated the export and import of goods and services in a cost-effective manner. Moreover, the development of containerization, large-scale tankers, more fuel-efficient jumbo jets, and air cargo transportation have collectively reduced transportation costs.

Even so, many global industries remain highly concentrated in a few countries, and often in a few narrow localities within these countries. Paul Krugman has advanced a framework to explain why industries cluster in particular locations. This framework identifies three factors — strong economies of scale, strong forward and backward linkages, and lower transportation costs. His thesis suggests that even with widening globalization, production may be even more concentrated because of the above factors.[10]

Global Economies of Scale and Scope

Global economies of scale apply when single country markets are not large enough to allow competitors to achieve optimum scale. Scale at a given location can then be increased through participation in multiple markets, coupled with product standardization. In the electronics industry, size has become a major asset. The Japanese and Korean electronics manufacturers have built their global strategy on the basis of scale and scope economies. Thomson, the French electronics maker, realized that a worldwide presence had become an imperative and proceeded to acquire General Electric's RCA consumer electronics business.

In many cases, it is the global economies of scope (gains from spreading activities across multiple product lines or businesses) rather than economies of scale that compel businesses to internationalize. Economies can be present in every function of the business, including manufacturing, purchasing, marketing, research and development, service networks, sale force utilization, and distribution. Such linkages provided the path for Proctor and Gamble to leverage its product strengths in Europe to more culturally-distant markets such as Japan. Similarly, Uniliver and Colgate-Palmolive enjoy the benefits accrued from multiple sources of consumer research, product development, and marketing and distribution activities.

Trend Toward Global Sourcing Efficiencies

The competitive nature of worldwide business has prompted companies to seek new sources of raw materials and components. Within domestic borders, companies continually examine the quality, cost, and availability of their domestic sources. More often than not, the decision to move offshore is a result of a significant change in one or more of these factors. For example, it is acknowledged that U.S. firms have moved operations abroad to sustain their competitiveness. Dramatic examples are seen in the rapid rise of Japan and Korea in the

1960s and 1970s due to their lower labor costs. As these countries have prospered, so have their wage costs. Consequently, operations have continued in low-skilled, labor intensive countries such as Thailand, Malaysia, Indonesia, Vietnam, Brazil, Mexico, and the Eastern European countries.

However, lower costs alone, is an insufficient reason for global sourcing. The people in these countries should have the requisite skills for producing the products. In semiconductors and consumer electronics, some of the labor-intensive activities are performed much more efficiently in these countries. Thus, global sourcing can provide a similar effect as economies of scale on the threat and rivalry among competitors. Moreover, marketing-intensive consumer products businesses like disposable razors are still the preserve of American and European producers.[11]

Falling Tariffs, Customs, and Taxes

National governments have played a prominent role in protectionist trade restraints. This role takes the form of enacting tariffs, non-tariff barriers, restrictions on imports and exports, custom duties and taxes. Their focus is protectionist (sheltering an infant industry from undue competition) and revenue enhancing (proceeds as major sources of revenue). In general, government policies can greatly influence the form and extent of global participation as in the case of the media industry, where, in many countries, foreign control of the media is restricted or prohibited.

In principle, there is agreement that free trade among all nations would lead to increased global competitiveness that would, in turn, result in higher quality, lower priced, and globally available products. The General Agreement on Tariffs and Trade (GATT) was established after World War II precisely to encourage free trade among nations through the systematic removal of trade barriers. GATT's 99 member countries have succeeded in reducing tariff revenues from a high of over 40 percent of government revenues in the 1940s to only about 5 percent, and under the Uruguay Agreement, they will approach 3.9 percent in 2000.[12]

While this development may be encouraging, the future of multilateral trade agreements remains highly uncertain. The reduction of tariffs has moved in parallel with the imposition of non-tariff barriers, rendering GATT powerless to handle. Non-tariff barriers include import quotas, voluntary export restraints, local content requirements, local technical standards, preferred government procurement policies, export subsidies and incentives, and even predatory "pricing." In addition, the decision by nations to engage in bilateral negotiations, and the emergence of regional trading blocs such as the EEU and NAFTA have undermined GATT's integrity and threatened its future.

Perhaps the most significant event is the launching of the WTO — the World Trade Organization — in January 1995, thereby ushering in a comprehensive multilateral trading system. This organization, in effect, replaces the General Agreement on Tariffs and Trade (GATT) which was established after World War II precisely to encourage free trade among nations through the systematic removal of trade barriers. The WTO held its first Ministerial Conference in Singapore in December 1996. This marks the recognition of Asia as a powerhouse of international trade and commerce.

While the establishment of the multilateral system characterizes the new trade order of the 1990s, there is a rise in regionalism, and of regional trading blocs in particular. This is manifested in the surge in the number of regional trade agreements formed in the 1990s, the expansion of the European Union, and the discussions regarding the creation of a Free Trade Area of the Americas.

Whatever direction this may take, it is clear that government policies are a major factor in determining whether an industry can realize its globalization potential. The easing of governmental restrictions can set off a rush for expanded market participation. Favorable trader policies can also break down entry barriers and increase rivalry among existing players. Government action can impede or facilitate the promotion of compatible technical standards. Compatible technical standards can trigger globalization and pave the way for the global dispersion of products. In summary, government actions are highly significant and should be monitored carefully.

How Will Globalization Affect U.S. Corporations?

Futurists have long been speaking of a global village, and this trend is well developed. Any firm that hopes to compete in today's world must work rapidly and across borders. Competition is global, but so are alliances and so is opportunity. Almost every major competitor sells its products in multinational markets and taps local capabilities. Even tiny four or five person software firms often offer foreign language versions of their products.

A microwave oven built in Mexico could be sold in Brazil or London under a variety of local brand names. A computer built in Singapore by a U.S. firm might end up being sold in Europe with a British brand name. Rapid cycle firms do what they must to offer new products quickly at attractive prices.

Box 3-2

> **"It does not matter whether a cat is black or white, as long as it catches mice."**
>
> Deng Xiaoping

One interesting sequence of events occurred in the early 90s, and it startled Washington by how quickly Information Age firms adapt to interference. This marked one major battle in the ferocious notebook computer wars.

The big U.S. companies were slow to produce portable PCs, and the Japanese, inevitably, seized the opportunity. Their banner competitor was Toshiba, which did its customary good job. Their first model, the T1100, came out in 1986, and they moved into a blistering cycle of proliferation and niche marketing. By 1990 they had the "luggable computer" market covered from 80C86s with 20 Mbyte disks at $1,500 to 80386s with 200 Mbytes at $10,000. Toshiba was a solid number one, and the market was growing at 35%.

A familiar story, perhaps, but in today's world of fast-moving skunk works and rapid-cycle core teams, things are not always as they appear. In 1991, Toshiba America laid off 250 employees, profits plummeted, the top manager of the computer systems division resigned, and they fired their distributor. Toshiba America in 1989 had asked its Tokyo HQ to design an 80286 *notebook* with a hard drive, but were told it was impossible. Tokyo was wrong.

COMPAQ announced its LTE 286 in October, and nightmare began. Toshiba's parry didn't get to the market until February 1990, and it was substantially larger and slightly heavier. By the time Toshiba got an 80386 notebook, its rivals had faster models selling for up to $2,000 less. Some 130 companies were in the market, with COMPAQ in the lead and running at warp speed. Toshiba was a product generation behind in notebook computers, and had little apparent hope of catching up.

The U.S. was winning. Then our government decided to "help".

In August 1991, the U.S. government decided to enforce anti-dumping in flat panels. There were a few tiny U.S. firms with names like Planar and Magnascreen. They collectively employed a few hundred, and mostly made specialty plasma and EL displays for the military. They were an insignificant presence in commercial markets. Commercial laptop computers use LCDs, and the *only* source was Japan.

The U.S. government ignored horrified objections from U.S. computer manufacturers. They said the law protected *flat panel* producers from dumping, but it "did not allow them to consider potential injury to U.S. consumers or industry when making decisions". The law applied whether or not it helped, and they *agreed* enforcement "won't guarantee any success for U.S. companies in the marketplace". They said, "Whether the (flat panel) industry can advance is beyond our jurisdiction".

A 62.67% duty was placed on imports of flat panels, which represented about one third of the laptop's cost. IBM's Michael Dutton complained this was "an eviction notice from the U.S. government to the fastest growing segment of the U.S. computer industry".

Indeed. For the privilege of building a laptop computer in the U.S., the new law demanded an extra duty of about $800. No one could afford that.

The cabal — not the usual suspects, but tiny and obscure U.S. firms — expected to profit from their crafty lobbying. They probably expected that, although their products were hopelessly noncompetitive in commercial markets, perhaps some foreign component firms and U.S. computer firms would pay them in return for help in dodging tariffs.

They probably expected fees, subsidies, and cheap technology access in exchange for wrapping their firms' names around others' technology and products. That is not what happened.

What caught the government by surprise was that the U.S. computer vendors saw no advantage in dealing with a group of parasites whose only asset stemmed from a bit of tricky lawyering. To the best of our knowledge, none of the over 100 affected companies made business deals with the cabal.[13]

Within days, COMPAQ, TI, IBM, and Apple had moved their notebook computer assembly offshore. Even then tiny In-Focus, an Oregon manufacturer of projection display products, dodged the bullet. They shipped their products to Canada to have the displays inserted.

This disruption cost the U.S. at least 50,000 jobs. A few CEOs visited Congress asking for intervention to save U.S. jobs, but they failed. It was hopeless. Not only did Congress not care where the jobs were, they suggested doing computer assembly in Mexico.

The computer industry, quite used to fast cycles and unexpected discontinuities, did what they must to survive. They protected their cycle time advantage by crossing borders. Rather than concede Toshiba and other arch rivals a cost advantage or time to catch up, they immediately shifted production. Despite blundering government intervention, the U.S. remains competitive in notebook computers.

In the end, the onerous flat panel tariffs were quietly repealed in 1993. The only lasting legacy was increased distrust and a deepened conviction that the U.S. government does not understand the basis for competitive advantage in the Information Age.

The tale of how and why U.S. firms declined to invest substantially in LCD flat panels, an obvious opportunity, is another story that is perhaps worthy of a book in its own right. Japan now controls 95% of this very lucrative market, worth $3.8 billion in 1992, $8.4 billion in 1995, and $15.2 billion in 2000.

The industry required decades of losses and such massive investments (plants alone cost $1 billion) that run rates of over a billion dollars per year are required for pay back. These have finally been achieved. Sharp's LCD sales for the year ending March 1994 are estimated to be about $1.6 billion. Seiko-Epson and Toshiba are right behind with about $500–700 million, and there is a pack of suppliers (NEC, Casio, Hitachi, Sanyo) in the $300–400 million range. (Business Week 1/17/93)

How Well-Prepared are We for the Global Age?

The educational levels in the U.S. are frighteningly low, and still dropping. A four-year Department of Education study gave 35–40 tests to some 26,000 people who were a cross section of the population.[14] Their 150-page 1993 report concluded about half the U.S. workforce over age 16 is unfit for any jobs requiring basic reading and writing.

The trend is disturbing. A study in the late 80's claimed a 37% functional illiteracy rate in the U.S. workforce. SAT scores dropped 76 points between 1960 and 1990.

Ranking the U.S. against other industrialized nations yields a disturbing pattern. For scientists and technicians per 1,000 people, we are #29 of 30. For books published per 1,000 people, we are #26 of 26. For the longest school year in days, we are #15 of 16. For population percentage covered by health care, we are #19 of 19. For growth of exports (nine-year average) we are #19 of 19. For science test scores (14-year olds) we are #15 of 17.

We do lead in some areas: We are #1 of 19 in infant mortality rate, #1 of 20 in divorce rate, and #1 of 19 in teen pregnancy rate. We are #1 of 17 in murders, #1 of 18 in rapes, and #1 of 16 in drug offenders per 100,000 people. And we are #1 of 14 in CEO compensation, both in absolute terms and in relationship to average workers' pay.

One encouraging statistic is that we lead in PC usage, with about 31% of homes having computers. We have tools, but we lack training.

There was a recent $2 million study of education of 175,000 nine and thirteen year olds in 20 countries. Germany and Japan declined to participate. Singapore did not participate either, and they have been beating Japan recently.

In mathematics and science, the results were consistent. South Korea, Taiwan, and Switzerland finished 1, 2, 3 in both categories. Hungary and the former U.S.S.R. were close, finishing 4, 5 in Math and 5, 4 in Science. Canada was 9th in both. The U.S. ranked 16th in both.

In math, we tied with Spain, and were just ahead of Jordan. In science, Spain was ahead of the U.S. We tied with Scotland and we were ahead of Ireland. This low ranking is a cause for concern. The low average wages and the high unemployment levels in Spain or Ireland are near disasters.

It won't get better soon. U.S. secondary school children work 1,000 hours per year. Japanese children put in 1,600 hours per year. At the senior high school level, the average U.S. student does 3.8 hours of homework per week, but the Japanese student does 19.0 hours.

American adults did spend more time with their families and less time watching TV than their Japanese equivalent (16 hours versus seven, and 12 versus 20). There have also been several studies that we spend at or near the top of the industrialized world in the cost to educate K-12 students, though we are #8 of 8 nations in what we pay teachers. (A good tabulation of such data is Shapiro's *We're Number One*, by Vintage Books.)

The New Competitors

In his book, *Competitive Advantage,* Michael Porter discussed why certain industries are more intensely competitive than others.[15] For the most part, his reasons mirror classical microeconomic theories: slower growth rates, higher fixed costs, the lack of product differentiation, and the absence of a price leader. Yet, he also introduced another factor: *mavericks,* or competitors that do not play the usual rules-of-the-game.

The Japanese are regarded as mavericks in today's global competition. Emulation of them has led to the birth of the four capitalistic dragons: Korea, Taiwan, Hong Kong, and Singapore. Even so, their arrival into the world scene has not been met with great fervor. In his best-selling book, *Rising Sun,* Michael Crichton describes the surge of Japanese multinational companies in America as "being disemboweled."[16] Robert Reich and Eric Mankin caution against joint ventures with Japanese firms for fear they are "a part of a continuing, implicit Japanese strategy to keep higher paying, higher value jobs in Japan and to gain the project engineering and production process skills that underlie success."[17]

While there is no disagreement about Japan's status as a world economic superpower, there is less agreement on how the Japanese have succeeded, how they sustain their competitiveness, and, consequently, how to compete or collaborate with them. For most academicians and business practitioners in the Western world, the success of the Japanese economy defies precise theoretical categorization. This is evident from the variety of explanations attributed to the success of the Japanese economy over the years. Earlier explanations focused on lower Japanese labor costs relative to those of more advanced economies.[18] When Japanese wages eventually surpassed competitor wages in the early 1970s, a new explanation was needed: a genre of books that extolled the virtues of Japanese management, notably Richard Pascale and Anthony Athos' *The Art of Japanese Management* and William Ouchi's *Theory Z,* quickly became best sellers.[19] Heightened comparisons between government-business relations in both Japan and the United States in recessionary periods of the 1980s ushered in another explanation: the success of Japanese industrial policies, as defined and guided by Japanese bureaucrats. Current populist explanations suggest that it is the Japanese industrial structure (*i.e., keiretsu*) that endows Japanese firms with significant competitive advantages in international competition.

In the wake of Japanese successes, Western observers are prone to explaining their success in ways that accommodate conventional theories and beliefs: lower wages, favorable exchange rates, generous government subsidies, and, at times, even unethical behavior. Lester Thurow flatly suggested that the example of the Japanese contradicts conventional economic theories.[20] What is different about the Japanese, or the Koreans, that make them formidable competitors and mavericks in the world market?

Synergies Derived from Business-Government Relations

While most Japanese academic scholars believe in the tenets of the market, they do not necessarily subscribe to its centrality — a viewpoint that might explain why government officials and bureaucrats have historically intervened in the private sector with the expressed aim of achieving particular ends.[21] Berkeley professor Chalmers Johnson suggests that government and business are engaged in power struggles and "turf wars" among themselves, but cooperate with each other where their goals coincide.[22] Under this interpretation, government, through use of policy instruments, "indicates" its preferences through non-coercive methods. Such "indications," however, must be formulated in light of the private sector's ability to thwart them. Private industry is free to act on its own. If it chooses to go against governmental wishes, it does so at the potential loss of special governmental consideration.

Post-war Japanese government actions has been to reward the strongest firms in industries that have proven their grown international strength and resiliency. Some of the early "winners" were chemicals, machinery, nonferrous metals, utilities, and iron. In the 1980s, government selections include high technology, industrial robotics, and artificial intelligence. Industries are generally selected on their capability to compete in the world market; their ability to benefit from economies of scale that can take advantage of the latest technology; and their capacity to offer good employment opportunities in the domestic market.

At present, the Ministry of International Trade and Industry, while still very influential, has a diminished role in targeting and administrative guidance. New government borrowing has created new financial markets. More successful firms have reduced their debt leverage, making them less vulnerable to policies of Japan's central government and ministries. Even so, it is hardly questionable that both MITI and the Ministry of Finance exert tremendous influence on the affairs of business.[23]

Synergies from Interfirm Collaboration

History shows that collaborative structures — specifically between banking and industry — can redound to the economic success of a country. This is true for England, Germany, China, Japan, and Korea. When government, business, and banking share risks, then the result is often vigorous economic growth. On the surface, the relationship between business and banking in Japan (i.e., industrial structures) appear similar to those of Venezuela and West Germany. However, unlike those groups in other countries, Japanese groups have distinct features that lead to competitive advantages. This is true for Korean industrial structures as well. While the features of these groups are becoming better known, they are not as well understood by most foreigners.[24]

Prior to World War II, the industrial groups in Japan were closely-knit family enterprises call "zaibatsu" — holding companies built around a bank, trust, or insurance company. Despite post-war policies to dissolve these groups, the groupings persisted and eventually more diversified and loosely-coupled enterprises — referred to as *kinyo keiretsu* (financial lineage) and *kigyo shodan* (enterprise group) reappeared. Among the more prominent in these groups were Mitsui, Mitsubishi, and Sumitomo. Others included Fuji, Daiichi, and Sanwa, and two other types of industrial groups in Japan, the bank group and the industrial family. The bank group consisted of a major bank and the large companies that are dependent on it for funds. By contrast, the industrial family comprised a major firm, its subsidiaries and its primary, secondary and tertiary subcontractors. Most of the *keiretsu* groups have a trading company to procure raw materials, provide market information, and manage distribution channels.[25]

These industrial groups share four common features that make them distinctive institutional forces within Japan. Mitsui *keiretsu* provides a point of reference of our discussion. First, firms within the Mitsui *keiretsu* share in a sense of community in that they clearly identify themselves and are identified by others as belonging to a particular group. Second, firms within the group are hierarchically organized, primarily along the lines of company size. Third, members within the group all tend to specialize in a given industry or, within an industry, a specific product or portion of the production process. Fourth, member firms hold shares in each other firms as an expression of the relationship that exists between them.

Given the inadequacy of capital/equity markets during the post-war period, the groups assumed the role of internal capital markets. Larger-sized firms were able to borrow funds from the Bank of Japan, for example, for which some were funneled to smaller-sized firms within the *keiretsu*. While there is fierce competition between the different keiretsu, the relationships among the firms in the *keiretsu* tend to be cooperative, if not patronizing. Officials from these different organizations meet in monthly sessions to share information and discuss matters of interest. Suppliers within the group make every effort to make timely deliveries to their clients. At times, suppliers may even go outside to other *keiretsu* to obtain the supplies to fulfill their commitments for deliveries.

Industrial groups such as the *keiretsu* offer tangible, economic advantages. U.S. automobile firms need to manufacture part of their supplies (tapered integration) partly to control their suppliers. Even with tapered integration, however, U.S. automobile firms do not enjoy the stability of deliveries. Nor are they assured of the certainty of prices as compared to their Japanese counterparts. Moreover, transaction and information-gathering costs are lower in Japan, since cooperative mechanisms exist within the industrial groups. Frequently, market surveys and reports, prepared by trading companies, are shared by members within the *keiretsu*. Already, these types of cooperative efforts have placed U.S. high technology manufacturers, that prefer to perform these activities individually at some advantage.

Synergies from Cooperative Management Systems

There is no dearth of materials that compare management systems of different countries. However, there are very few attempts to explain how differences between management systems translate into specific comparative advantages. In this section, we will summarize some findings and observations regarding the differences between Japanese and Korean management systems, that are based on Confucianism, with those of the United States. Then, we attempt to link these differences to the broader context of industrial policymaking.

Ouchi described Japanese organizations in terms as having lifetime employment; slow evaluations and promotions, non-specialized career paths, collective decision making and responsibility, and a wholistic concern for employees.[26] In contrast, U.S. organizations are characterized by short-term employment, rapid evaluation and promotions, specialized career paths, individual decision making and responsibility, and a segmented concern for individuals.

Box 3-3: Working Hours/Year:

Seoul-2,302
Bangkok-2,272
Hong Kong-2,222
Copenhagen-1,689
Frankfurt-1,725
London-1,880
Far Eastern Economic Review November 24, 1994

Two main points exemplify these differences. One is the Japanese notion of groups and groupism, and the second is their long-term outlook. To understand groups, one has to realize that Japan, about the size of the state of Montana, has about half the population of the United States (i.e., 119 million versus 234 million in 1983), in a smaller place that is not endowed with rich natural resources. As early as childhood, Japanese are then nurtured to be part of a group, to partake and share in the activities of the group. Historically, groupism was enhanced by a rather homogeneous population and a period of isolation in the Tokugawa era. Therefore, the loyalty and identification of an individual to a group, or a corporation, are paramount. In addition, the Japanese tend to adopt a longer time horizon, as reflected in the long period of job training (13–15 years), slow promotions, and long-term evaluations. Deep-seated loyalty to an organization and a sense of work discipline resulting from long years of training are common correlates of an individual Japanese employee.[27]

While the appropriateness of each of these systems has to be evaluated in terms of the larger culture in which they are imbedded, there are plausible indicators of relative effectiveness. Organizations in the U.S. often find them-selves in an "us" versus "them" relationship with their employees. Thus, there is a strong adversarial relationship between the parties that can result in prolonged collective bargaining sessions, if not in longer strike periods. While strikes occur in Japan, they are not as prolonged and expensive as those in the United States.

Box 3-4

> **Asians will look at those who do well and try to find out how it happened; they won't ask for a handout. Asians must depend on themselves and no one will bail them out. This is their biggest advantage.**
>
> Lee Kuan Yew, Senior Prime Minister, Singapore

Absenteeism and turnover statistics present another plausible dimension in which to compare systems. In many industries in the U.S., daily absence rates approach 15 to 20 percent and cost an estimated $26.4 billion annually. In addition, worker layoffs, job turnovers, and absenteeism have undermined the international competitiveness of U.S. firms.[28] More rigorous comparisons are not possible because comparative data are not available. Even so, the discipline of the Japanese workforce has been cited as reasons for the high quality of their export products. In contrast, many of the defects on U.S. manufactured automobiles have been traced to employee negligence.

Emerging Imperatives: Globalization and the Information Age

According to the Economist:[29]

> Over the years the pace of economic development seems to have quickened. The industrial revolution in the 18th and 19th centuries was a slow affair compared with growth rates today. Thanks to better communications, technology is now diffused more quickly than in the past. After the industrial revolution took hold in about 1780, Britain needed 58 years to double its real income per head; from 1839 America took 47 years to do

the same thing; starting in 1885, Japan took 34 years; South Korea managed it in just 11 years from 1966; and, more recently still, China has done it in less than 10 years.

People today have far more access to greater amounts of information than any time in world history. With a flick of a switch, we can see what is happening in other countries. This capability to transmit information instantly has multiplied business opportunity, exposed risk, and intensified reactions to managerial decisions.

Kenichi Ohmae, Japan's most respected management theorist, and author of a recent book, *The End of the Nation State*, suggests three accelerating effects information technology has had on globalization:[30]

- At a macroeconomic level, information technology has made it possible for capital to be shifted instantaneously anywhere in the world. The volume of finance trade is approximately $4 trillion a year. Foreign-exchange transactions amount to about $1 trillion a day, or $250 trillion a year. This is forty times the size of trade and services. The productivity of certain industries, such as telecommunications, have been greatly enhanced. In Hong Kong the average businessman is nearly as likely to have a cellular phone as a watch!
- At a company level, information technology has changed what managers can know in real time about their markets, products, and organizational processes.
- At a market level, information technology has changed what customers everywhere can know about the way other people live, about the products and services available to them, and about the relative value such offerings provide.

Services, services, and more services

The Information Age has accelerated the pace of industrial transformation from manufacturing to services. In the United States, manufacturing holds a steady 23 to 24 percent of GNP. This is true of all the other major industrial economies as well. Manufacturing *per se* is not yet in decline; it is employment in manufacturing that is declining.[31]

Still, the U.S. manufacturing sector is under an attack that goes beyond corporate downsizing. There is a reinforcing confluence of government economic policy initiatives (from employment and patent law, to trade agreements, to taxation and regulation) that could reduce this sector to under 10% of GNP. Of those who notice this trend, most assume that it is coincidental. However, some experts are starting to say that it may be an intentional U.S. government policy to expand beyond national to international wealth redistribution.[32]

Employment figures are even more staggering: about 80 percent of the workforce is now employed in services, contrasted to 17 percent in manufacturing and 3 percent in agriculture. By 2000, manufacturing employment is forecasted to drop from 17 to 12 percent, with the major gains accruing to services. A large part of this is caused by the fact that Information Age products tend to have extensive service components. Each time a corporate user upgrades to a new generation of computer, the training and support cost is between $5,000 and $10,000 — more than the new computer itself costs. *Services are part of the product; the product is part of the service.*

The fastest growing sector of world trade is trade in services. In the United States, services account for about 65 percent of GNP. Few know that the U.S. service balance has been increasing since 1983, close to offsetting our merchandise deficit. Japan, on the other hand, runs a service trade deficit.[33]

The bad news is that most of what we teach and/or practice are mired in assumptions rooted in manufacturing, not service, enterprises. This is more apparent in marketing and management classes.

One of the distinctive features of services is co-production. Services are produced and consumed simultaneously thereby involving the customer to some degree in the actual production and delivery of the service. Competitive advantage in the service provision often results from superior standard of service, rather than any purely physical attribute of the product concerned.[34]

When asked what transformed Hewlett Packard into a $40 billion giant, Lee S. Ting, General manager of Asia-Pacific Operations, said: "We have a great product (H-P printer line), but it does not stop there. Our sustainable advantage is service — we get over 10,000 service/information calls a week, and we have devised an excellent information support base for handling these calls. We do this better than everyone else. This is why we have been very successful."[35] For Hewlett Packard, success derives from its ability to integrate the requirements of global strategy with its provision of worldwide services supported by global information systems.

Asia: The Next Economic Frontier?

Two trends — globalization and the Information Age — have affected no region of the world any better than Asia. In real terms, Asia's exports (excluding Japan) are 13 times higher than they were 25 years ago.[36]

Despite the financial crisis, Asia's performance is worth noting.

Asia's share of world output tripled in the last three decades, close to 25% of global GDP. The region has maintained an average yearly growth of 7 percent since 1970. In the 1970s and 1980s, Japan led the way. In 1990, the world's total GNP was 23.4 trillion dollars. The U.S. accounted for 24%, the EU for 21%, and Japan for 13%. These ratios are not expected to change significantly by 2010, when world GNP is projected to reach 42 trillion dollars (at 1990 prices). The U.S. is expected to account for 21%, the EU for 24%, and Japan for 17%.[37]

In the mid-1980s, the picture shifted to the four tigers — South Korea, Singapore, Hong Kong, and Taiwan. In recent years, center stage in terms of the highest growth is directed at ASEAN; today, it is directed at coastal China.

Asia offers economic and market opportunities for a host of industries. China is increasingly recognized as the next world's growth center. The ASEAN countries have surged ahead, and the coastal region of China is simply exploding. Note that the per capita GNP in China is a mere $317, but that Shenzhen, close to Hong Kong, boasts a per capita GNP of $5,695.

Not unexpectedly, wealth has its bedfellows. Consider the income generational gap that pervades the fast growth Asian economies. The late Malaysian tycoon Tai Chik Sen, whose family controls the big Soon Seng Group, maintained a humble lifestyle even after accumulating tremendous wealth. When local reporters asked about his modest lifestyle in contrast to his sons, who were driving around Kuala Lumpur in flashy BMWs, he said: "My sons have a rich father. I don't."

A big barrier to Asian growth and development into a competitive regional bloc is the disparities in the levels of development from country to country. Economic disparities among Asian countries are greater than among countries in the EU or NAFTA.[38]

Asia's most advanced nation is Japan, a G-5 member with a per capita GDP of $40,000. Following Japan are the four newly industrializing economies — Korea, Taiwan, Singapore, and Hong Kong — where per capita GDP ranges from $10,000 to above $20,000. The third highest group of countries includes ASEAN countries and China. The lowest income group includes Cambodia, Laos, and Myanmar, the so-called "semi-ASEAN" nations with per capita GDP of only about $200.[39]

The Rise of Asian Networks and Region-States

A special feature of the business structures of the economies of Asia is the structure and function of the overseas Chinese. It is as much a phenomenon as the Korean *chaebol*, and the Japanese *keiretsu*.[40]

The Chinese business networks are largely personal networks and are organized primarily through kinship circles and ties of common origins. The Japanese business networks are predominantly intercorporate ties, also called *nemawashi*. The South Korean chaebol are dominated by elite business families that are privileged by a strong state.[41]

The dynamics of growth in Asia cannot be explained without reference to the Overseas Chinese — a group, now numbering about 60 million, and with assets at about $2 to $3 trillion. John Naisbitt calls them the "greatest entrepreneurs in the world."[42]

According to Naisbitt, Fujitsu Research in Tokyo looked at listed companies in just six Asian countries. The overwhelming majority were owned by Overseas Chinese:

- Thailand 81 percent
- Singapore 81 percent
- Indonesia 73 percent
- Malaysia 61 percent
- Philippines 50 percent

Already, the economy of the borderless Overseas Chinese is the third largest in the entire world, only outranked by the United States and Japan. As the dynamism of these entrepreneurs continue, we will soon see the world moving from a collection of nation-states to a collection of networks.

Implications for U.S. Firms

U.S. firms operate in a different competitive environment for which they have derived competitive strengths and also suffered comparative disadvantages. The major distinction to be made in regard to the government policy of the U.S. as opposed to Japan can be made along the lines of reasons for government intervention into the competitive market. The U.S. intervenes mainly for purposes of regulation. This is typified best by antitrust legislation and the FTC's veto power over mergers. Both represent interventions which regulate or police competitive activity rather than develop it.

By contrast, government interventions, as depicted in Asia, notably Japan, are focused more on the development of the national economy and a recognition, subsequently, of the need to shape competitive activity so that development in a particular direction will occur. This "regulatory state vs. developmental state" dichotomy suggests specific differences in the orientation of competitive strategies between the U.S. and Japan. This distinction is accommodated and sustained by the institutions of the two countries. For example, Japanese have tended to have a global and long-term outlook. This is particularly true in Japan since it has always been dependent on other countries for its raw materials.

Given the size of Japan's domestic market, it has become crucial to export heavily in order to compensate for its strong reliance on imports. The outlook in the United States has been primarily domestic in orientation. Given its rich natural resources and the vast size of the domestic market, it was expected that firms meet the needs of their domestic market before supplying foreign markets.

In regard to the Japanese *keiretsu* or the Korean *chaebol,* there are no comparable industrial groupings in the United States.[43] Instead, industry groupings such as the SIC (Standard Industry Classifications) which are used primarily as reference points for the extent of industry concentration. Even where some loose groupings exist, these do not approximate the amount of control and interfirm transactions that occur with Japan's *keiretsu*. The reason is institutional: in Japan, as well as in Korea, the groups served the purpose of internal capital markets. In the U.S. where there is a vast amount of equity, the need for large conglomerates to function as credit-houses never materialized.

Other contrasting institutional features include the nature and importance of bureaucracies in each country. Government bureaucracies in Japan are held in higher esteem than those in the United States. Ezra Vogel has noted that bureaucrats in the MITI come from highly regarded Todai.[44] Solidarity and esprit de corps among them develop from their school days as well as their years of training in MITI. The relative influence of the business world on government policies is quite strong. Not unexpectedly, many of Japan's prime ministers have come from the ranks of MITI.

In the U.S., the bureaucracy is suspect and the influence of particular business groups, i.e., Roundtable, is limited. In fact, one of the strongest arguments against a national industry policy in the U.S. has been that previous government decision to support some industries (trucking, railroads) have not been encouraging.[45]

There are many who have argued that U.S. government policies, particularly those dealing with antitrust regulations, have not been conducive to economic development.[46] The fundamental issue revolves around whether or not the government should institute commercial policies. All governments do. The issue is whether or not those commercial policies will be coordinated and directed toward a national economic goal. The regulatory nature of most U.S. policies suggests that the government is struggling with the former issue and not the more important latter.

There are other implications for U.S. firms in regard to the Overseas Chinese as a network of networks. They know each other, and they work together when necessary. They are intensely competitive but generally trust kinship and not those outside the same family, village, or clan. One Hong Kong banker is cited: "If you are being considered for a new partnership, a personal reference from a respected member of the Chinese business community is worth more than any amount of money you can throw on the table." The origins of the Overseas Chinese, a predominant majority came from one place: China's Fujian Province (the other is Han). Even Singapore's Lee Kuan Yew's ancestry hails from Fujian.[47]

Finally, a belief that is gaining currency is that globalization presents both opportunities and limitations for participating in markets outside the home country. To compete effectively in the new environment, firms will have to develop a global mindset. For many years, this belief was given lip service, but not seriously entertained. External events in the world suggest that a new set of realities have been emerging, and significantly impacting how companies need to be managed to ensure success. It is no longer conceivable that any company, regardless of size, can continue to be successful by confining itself to its domestic market, however large that market may be. Competing in a new global environment is one of many significant challenges faced by the contemporary manager.

Our scorecard on our current attitudes about doing business abroad, as well as beliefs on the effectiveness of our educational systems in preparing us for a global world reveals deep gaps. In some fundamental ways, U.S. firms are relatively inexperienced in terms of competing globally — a feature that favors the Europeans or the Japanese who have had to compete on this basis for several decades. A brief survey of our educational needs suggests that we are slipping further behind in understanding the essentials for competing globally. And, finally, even the lure of cross-border alliances, heralded as the vehicle for achieving rapid globalization, is limited in that there is more to be learned about cooperative partnerships. To meet the globalization challenge, it is evident that Old Think premises have to be discarded. New ways of conceiving global competition are needed.

Chapter 4

The New Technology

Key Themes at a Glance

The Western world, particularly the United States, gained ascendancy and earned its place at the forefront of world economies based on a mastery of Machine Age technology. The "sleeping giant" awoke when bombs fell on Pearl Harbor, and the upsurge of industrial production led to dominance based on technologies ranging from aircraft, to jet engines, to computers and rockets with nuclear warheads.

When Machine Age met the Information Age head-on during the Gulf War, it was no contest. Japanese chips, along with commercial laptop computers, logistics databases, videogame technology, and GPS systems gave Western troops battlefield advantage. Modern communications, smart weapons, and small quick teams let Western commanders, for once freed of Machine Age control themselves, easily rip apart Iraq's centralized command and control structures and negate or exploit its Army's preprogrammed responses.

More significantly, the drivers for new technology have shifted from government "mega-programxs" to the commercial sector. The once trend-setting Western Military-Industrial complex has become a backwater. Netscapes' young wizards left the government sector because they wanted to work with more

advanced technology. Today's PCs exceed the power of Cold War era supercomputers. Computer graphics is now driven by movie making, not nuclear weapons design. Cryptographic technology is migrating to mass market PC applications. The Internet, developed to allow survivability during a nuclear attack, now offers hope for truly global commerce, free from Megastate bureaucrats and national boundaries.

This chapter examines Information Age technology. What has changed? How are the new technologies different? Why does Information Age technology necessitate new management viewpoints, approaches and business models? And, the key question, how can business exploit Information Age technology to achieve better results on a sustainable basis?

A Wake Up Call

In 1985, the editors of *Business Week* sounded a belated alarm: America's high technology industries were in crisis![1] At that time, high technology was considered the shining pinnacle of America's economy, its citadel, its bastion. While it was widely acknowledged that textiles, steel, machinery, automobiles, and consumer electronics had fallen prey to foreign competitors, conventional wisdom held that U.S. high technology industries — semiconductors, computers, telecommunications, and biotechnology — were impervious to foreign competition.

After all, the transistor, the integrated circuit, the computer — products that had spurred the high technology revolution — were all invented in the United States. Moreover, America's scientists were the technological leaders, and U.S. high technology industries were the vanguards of future innovation.

In a matter of years, such optimism had collapsed. *Business Week* editors detailed how Japanese semiconductor firms had broken America's stranglehold in the 64 kilobit dynamic random access memory chips, heralding their emergence as formidable competitors in the high technology race. Since the 64k DRAM had been regarded to be a crucial input to a host of intermediate products, the loss of this market to the Japanese was momentous — a loss approaching crisis proportions. For many observers of this high technology battle, the unthinkable had, in fact, occurred: the Japanese had succeeded in semiconductors, employing similar entry-strategies previously used in penetrating the textiles, machinery, consumer electronics, steel, and automobile industries.

Ironically, the Japanese themselves had publicly claimed victory, years earlier and in one of our own publications, the famous October 1981 issue of *Scientific American*. How did they know that they had won years before we even noticed our lead was at risk? They knew because they understood the dynamics of steep learning curves, one of the Engines of Prosperity. We missed noticing this event. In fact, many Western leaders still remain oblivious to its lessons and significance — not because the information is unavailable, but because they are blinded by their own myopia.[2]

The aftermath of losing the 64k DRAM market left U.S. analysts perplexed as to how the Japanese had done it, and what appropriate strategies to pursue as a response. The fact that conventional economics said that Japanese semiconductor firms should not be able to recoup their huge initial investments and sustained losses bewildered the analysts. Despite the warnings by the editors and actions taken by the U.S. semiconductor industry, a similar event would occur several years later with the next generation of memory-circuitry, the 256 k DRAM, and again with 1 M DRAM, and again with larger sizes.

Korean and other firms used similar tactics to capture portions of this market. U.S. firms now generally cede the technology-driving memory chip market to foreign suppliers, while cushioning the blow by retreating to more complex circuitry in which they still retain some competitive advantages. Motorola, the last major holdout, gave up on DRAMs in 1997.

World leadership has been inextricably linked to technological leadership for over 200 years. Therefore, it seems peculiar that U.S. policy makers and corporate leaders don't evidence more concern about catastrophic repercussions should the U.S. relinquish its leadership in high technology. The fact that they don't testifies to some of the organizational learning disabilities that Senge and others have noted.[3]

Leaders and Losers

In recent years, the technologies with the biggest mass market commercial impact have all been invented in the West, from FAX machines, to compact disks, to VCRs, and videocameras. Without exception, the major revenues from these technologies have accrued to foreign competitors, mostly the Japanese. How can this be?

There are two possibilities. Either there must be an inordinate number of unlucky or incompetent people in critical positions at major firms, or something is very wrong with the models we use for making decisions and converting technology into business or societal results. We suspect the latter.

Consider what are arguably two of the biggest blunders made in the history of the computer industry. If John Sculley at Apple had, in the mid-80s, exploited the Mac's graphical operating system by adapting it to the Intel based computers that owned most of the market, Apple would today equal Apple plus Microsoft. If John Aikers at IBM had moved more aggressively into personal computers and related software, instead of clinging to big iron mainframes, IBM would be the sum of all the PC vendors, plus most of Microsoft.

Neither CEO took such action, and eventually, presumably as a result of these and similar decisions, both lost their jobs. Today Apple is moribund and IBM is a pale reflection of its legendary greatness. Were these bad decisions due to personal failings or some dysfunction in the system? Again, we suspect the latter.

These cases were archetypal of what has been happening at hundreds of firms. Both Aikers and Sculley were fairly typical Machine Age CEOs, who, as such, saw their masters to be Wall Street, and their jobs as adjusting business operations and sales to optimize profit flow, and, hence, to maintain stock price. We are convinced that virtually all Machine Age CEOs would have taken the same viewpoint. Indeed many other CEOs at large firms, from Robert Allen at AT&T to Ken Olson at DEC, probably suffered more public embarrassment from their poor technology decisions than the two discussed.

Box 4-1

"There is no reason anyone would want a computer in their home." Ken Olson, President, Chairman and Founder Digital Equipment Corporation, 1977

Conversely, the most successful Information Age CEOs, from Bill Gates and Craig McCaw to Andy Grove and Jim Clark, have all, at times, jumped to exploit new Information Age technology and trends, and hence to couple their firms to the Engines of Prosperity. They have on a consistent basis done this almost regardless of the impact on their near term balance sheets. All these firms have good records of consistent innovation and new product flow, and they have often done this at the price of occasional "failure" and temporary loss of profits.

McCaw Cellular was never profitable, but was sold to AT&T for billions of dollars. Gates invested in Windows without any financial return for ten years. It is an industry joke that — from Windows to Word to Powerpoint to Excel to Internet Explorer — Microsoft has to do a product at least three times before it gets one that is successful. Jim Clark started Netscape in business by giving away its products. In the early 80s, when most firms were laying off engineers, Grove increased Intel's new product development efforts. Intel came out of the recession with the 386 chip, resulting in eventual industry domination.

If New Technology is the Answer, How Can You Get Some?

One of the major problems Machine Age CEOs face is that there have been two established methods for developing new technology, the "pipeline model" and the "incremental model," and neither are delivering results. The problem is that neither model is, of itself, adequate in the new era. Let's examine why.

The pipeline model is sequential: first you did research, then applied research, then advanced development, then engineering, then manufacturing, etc. The old paradigms say that the source of wealth was technology creation. "Invent a better mousetrap, and the world will beat a path to your door." If this ever worked (we doubt it), it certainly has not of late.

Consider the old XEROX Palo Alto Research Center (PARC). It developed most of the technologies used in computers today, but was totally unable to convert them to commercial revenue streams to benefit XEROX. Much the same can be said for most research labs associated with Machine Age firms. Most of the new developments in communications have come from firms other than AT&T, even though most of the enabling technologies came from Bell Labs.

Another way to think of this is "supply side" technology. You did great research — like the transistor at Bell Labs or the new computer paradigms at PARC, and it generated wealth. The job of a CEO was to set aside some money, usually not a lot, and lock some "propeller heads" in a lab to write papers, win prizes, and do good things. Ever so often, a few of these things would eventually ("eventually" is a key word) make it out of the lab and into the world. The West was very good at technology creation, to the point where for the past few decades foreign firms were happily "donating" free technologists to work in U.S. labs.

Supply side technology is not a competitive model for the Information Age. Most of the time, trying to push technology on users just doesn't work. It's like pushing on a rope, an exercise in futility. The Government's lavishly funded "technology transfer" initiatives for "defense conversion" were dismal failures, to the point where they have been erased from political memories. Even when they want to, R&D labs lack the resources (production, practical engineering, purchasing, marketing, distribution, etc.) to convert technology into products and successfully introduce them by themselves. In large corporations, those who "own" these resources, the product divisions, sometimes would rather protect their turf than see a new technology succeed. Most division managers see corporate R&D as a "tax," and they usually prefer to keep the money themselves.[4]

Even in cases where all the disparate organizational functions can pull together to develop and implement a new technology, the pipeline model still has limitations. Mostly it is too slow, and mostly the researchers are too distant from the markets their corporations serve. In a few cases, Hewlett Packard for one, strong organizational culture and infrastructure continue to make this model a successful strategic resource for the corporation. Still, for the most part, corporations and the government are moving away from the pipeline model. Once mighty Bell Labs, a national treasure, no longer does basic research, only application.

The other model is incremental, where small teams do "insanely great products." This approach is a thing of legend, from Kelly Johnson's 1940s skunkworks at Lockheed to the early days at Apple, Tektronix, or a hundred other U.S. firms. Mostly, it doesn't work by itself anymore either.

One problem is that Machine Age firms don't tolerate rebels and mavericks, so skunkworks tend to be, sooner or later, stamped out or sabotaged and caused to fail.[5] Only the strongest and most knowledgeable upper management protection can preserve them, and most Western upper executives these days have financial, legal, or MBA backgrounds that are not rich enough in new product development expertise or experience. Another problem is that in today's fast cycle world of non-linear time (we'll say more about that in a moment), all the small teams are running flat out just to get extentional, next-of-type products done.

Pretty quickly, most of today's small team competitions devolve into "quicker, cheaper." The pace accelerates endlessly, and margins get very thin. Soon all the small teams are fully occupied doing the next products. Thus firms can be leapfrogged by rivals that invest in radically new technology, or undercut by firms that have less expensive corporate overheads, or preempted by those that are quicker.

Most computer users cared more about whether or not they could do their jobs better, than the fact that the OS/2 system was technically superior. Again and again, since users could only get the applications they wanted (from games, to spread sheets, to word processors, to Internet browsers) for Windows, that is what they chose. And so, OS/2, though brand new and technically superior, became a "legacy system" before it had any significant market impact. Bill Gates had used what we describe later as one "Engine of Prosperity" to lock his software in, and to lock IBM's out.

Concurrently, incremental efforts at both IBM and Apple were insufficient to hold a position in the mainstream of the computer industry. After a decade of improving Apple's Mac, it just was plain out of date as a platform. Slowly, even the most loyal users drifted away as the PC platform, which was evolving at a more rapid rate and declining in price faster, offered them better alternatives. Even when three major firms — IBM, Apple, and Motorola — teamed up to with an initiative to thwart Intel and Microsoft with their PowerPC, a major new technology platform update, it was pretty clear that the effort would be inadequate to break the platform "lock" that the PC had developed.[6]

Since both traditional methods for getting new technology were not working well, is it any wonder that most CEOs gave up? The smartest Machine Age CEOs were pretty sure that doing more of what they had in the past was not hopeful. They were also pretty sure that trying to change and develop more competitive new methods of commercial innovation was beyond them as well. Therefore, it was not unusual that executives like Norman Augustine, CEO of Lockheed-Martin, based his (successful) strategy to save his firm by buying up rivals, on the fact that no major defense firm had ever managed to succeed in commercial markets.[7]

We stress that this malaise is not idiosyncratic to defense contractors or high tech firms, it is very general. Not so long ago one of the most revered firms in the bicycle industry, Schwinn, a 97 year old Chicago corporation, declared bankruptcy. They did this in 1992, at a time when the industry was swimming in prosperity, a time when innovative and more expensive new bicycle designs were proliferating rapidly. Not coincidentally, the Apple and IBM crisis and CEO replacements we noted were both during times when technology and new product options were rich and the industry was prospering.

Why did Schwinn fail? According to *Forbes*, "Schwinn was obsessed with cutting costs, instead of innovation." Is this common? You bet it is. In a DPI survey of 200 Fortune 500 companies, two thirds said they had no formal manner to encourage the search for and the development of new products, customers, or markets.[8] Surrounded by huge lakes of market opportunity, and in times when new technology offers so many rich options, we are seeing an industrial die off because of an inability to adapt and innovate.

We submit that so many firms lack "innovation engines" because getting innovative new products quickly and consistently was just not that important in the Machine Age. Contrast that to the fact that the leaders in the most demanding markets today (yes, we are back to high tech) are almost universally paranoid about being early to identify and exploit new trends in markets and technology. Again, and we can't say this often enough: *The Machine Age was about repetition, the Information Age is about innovation.*

If the way to prosperity is technology, we see that the old models, pipeline and incremental innovation, are by themselves inadequate to be competitive in the Information Age. We see that most common "solution" to becoming competitive is cost cutting and buying rivals. Some still think that this will lead them to a prosperous future, but we disagree. Clearly, trying to shrink your way to leadership has fairly obvious limitations. There is compelling evidence that layoffs permanently damage a firm's ability to innovate.[9]

Putting Jet Engines on Tractors

The concept of putting a jet engine on a tractor is a silly example of inappropriate technology insertion. It would be very powerful, but also expensive and most of the power would be wasted. Still, at a major transition point, it is sometimes hard to say what is appropriate and what is not. If you look back to the early Machine Age, you find examples of tractors that were small versions of railroad locomotives, powered by steam. At some point, that option must have had seemed reasonable. To the best of our knowledge no one has seriously tried to put jet engines on tractors, but there were experiments by Detroit in the early 60s to insert them into cars.

Our Japanese friends are baffled by the fact that some Western firms are spending a lot of money to use what they call Flexible Manufacturing Systems (FMS) — we typically call this technology "robotics" — incorrectly. FMS, initiated by a Western (British) firm Molins Ltd. in 1967, is another example of an Information Age technology invented by the West and commercially exploited by foreign firms.

At the time of its creation, the Molins system was not cost effective. It therefore "failed." In effect, this technology allows miniature automated factories. Decades of work perfected the technology. It took more precise mechanical technology, microprocessors, robot technology, and the perfecting of quality control before this worked well. It also took markets where interfirm competition was fierce, markets were saturated, and standardized or commodity products were generally uncompetitive before this technology was useful. The Japanese deem that all these things have happened, and that, therefore, the time has come for flexible manufacturing.

We see it differently, apparently. Kodama cites a study that considered half of the systems installed in Japan and the United States. The study concludes that "For the most part Western Management has treated the FMS as if it was just another set of machines for high-volume, standardized production. Furthermore, in the United States, an FMS cannot run untended for a whole shift, is not integrated with the rest of the factory, and is less reliable." The average number of parts made by an FMS in the U.S. was 10, while in Japan it was 93. Conversely, the annual volume per part in the U.S. was 1,727, versus 258 in Japan.

So Japan makes about ten times the use of the "flexible" part of FMS as does the U.S. More tellingly, they use it to accelerate innovation and increase product differentiation. For every new part introduced into a U.S. FMS system, twenty-two new parts were introduced into Japan's. They cannot understand why we fail to correctly exploit this wonderful new technology.[10]

Reengineering, viewed by most as a new technology, has been one of the most popular fads to ever hit Western business, and this despite the fact that in the vast majority of cases it has not delivered the promised results.[11] Even the architects of reengineering confess that it has failed in most (70% of) cases.[12] One of them, Michael Hammer, did a fascinating article for *Wall Street Journal,* wherein he blames "the customers," not management and certainly not his methods, for the seemingly endless layoffs that U.S. industry is facing.[13]

One factor that is striking about the new technology is that Information Age computers, communications, and database technology and other forms of hardware and software can bring unprecedented freedom and empowerment to individuals or small groups. Consider, for example, that, at least in theory, the whole form of Western Democracy could be transformed and made more effective if this technology was properly exploited.

Many small firms, including The Trudel Group, have long based their business models on such exploitation of new technology. This chapter is being written some 2,000 miles from this author's home office. The two authors have been collaborating on this book electronically across physical distances that have ranged from between 100 miles (Eugene and Portland, Oregon) to almost half the planet, and from locations that have ranged from mountain cabins to hotels in Korea or Istanbul. Our book's publisher is in London, the editing and production is being done in Singapore, and the PR firm is in New York.

Technology allows vocational choice and non-traditional work environments. Indeed, the centralized factory settings and structured nature of Machine Age jobs may someday be viewed as historical anomaly. Most sales and product support jobs, and an increasing number of R&D jobs, are already "non-traditional." For example, a friend, laid off from his long employment as a top scientist for a large U.S. firm, now works for a Japanese owned California firm from his home in Oregon. He has company furnished workstations with an ISDN connection, and a full lab. By the year 2,000 most jobs will be "outside" the company. More than half (51.9%) of firms already use outside consultants on a regular basis during their new product development work.[14]

Box 4-2

"The age of the hierarchy is over."

James R. Houghton, Chairman and CEO, Corning

To get back to reengineering, there is also a dark side to Information Age technology, the grim, bleak vision of "big brother" and robot like workers that have been chillingly described in many works of fiction. This too is possible, since modern technology also allows processes to be redesigned, run, and monitored by computers. If the old Machine Age model was having the workers as cogs in a machine, now the cogs are computer designed and controlled. The "dark Satanic Mills" that the prophetic English poet William Blake spoke against some two centuries ago can today be found in many places and many nations.[15]

The new technology allows for individual keystrokes to be counted and managed, workers to be monitored and controlled by video technology, and for their e-mail and phone conversations to be recorded or monitored. In many firms this is already common practice. The Internet community cringed when the Clinton administration sought and received $500 million from Congress for widespread Information Age digital wiretaps. The technical ability to simultaneously tap one phone conversation out of 400 (1:100 in urban areas) is sought.[16]

The Wall Street Journal recently carried an article reporting that Southland Corporation, that owns the 7-Eleven franchises and is in turn owned by a Japanese firm, is doing exactly that with its franchises.

U.S. citizens, who spent their own money to buy a franchise and become independent small business people, are learning they only got the last half of that bargain. Some are enraged that they are expected to run their businesses and adjust their lives to serve the computers that run the stores. Mr. Toshifumi Uzuki, President of Ito-Yokada, which controls 7-Eleven, is determined that that is exactly what they will do. "American 7-Elevens don't have a choice: It's either adopt our plans or die," he said.[17]

"What does this have to do with reengineering?" you ask. Well, at least in theory, the basis for reengineering is scientific. The stated intent is to use the new Information Age technology to help corporations conduct business "better," more efficiently. In theory, reengineering and downsizing or rightsizing are different things. (We now have many euphemisms for firing people. Downsizing is when your colleagues lose their jobs. Rightsizing is when you or a family member are terminated, as has happened to one of three families in the U.S.[18] No one, of course, is responsible for "wrongsizing." It just happened.[19])

In theory, reengineering is done for companies to help them become more competitive. In practice, reengineering seems most often to be done to workers by outsiders. The field has been a gold mine for the large accounting firms, who are given sizable fees to come in, study, and re-engineer a company's processes. The result, if successful, is to allow work to be done faster and with fewer workers. Though there is distinction made between reengineering and downsizing, it is a distinction without a difference, at least so far as the laid off workers are concerned.

There is an old "I Love Lucy" TV episode, where Lucy is working in a candy factory. At first she copes, but as the Machine Age production line runs faster and faster she becomes further and further behind. She eventually loses all control, with candy going in all directions. It is a very humorous episode. The difference between reengineering and downsizing, we suppose, is that scientifically re-engineered Machine Age processes can be run at maximum speed indefinitely with the remaining workers.

This is exactly analogous to the FMS example, except that reengineering is now almost universal at Machine Age firms. It can be used not just for manufacturing, but for almost any process. It applies to sales, to order processing, to new product development, and, probably, though we have not seen documented cases, to how workers can most efficiently take lunch or bathroom breaks. It is a Machine Age manager's dream come true, for "how many, how much" can now be measured, quantified, and optimized with machine like precision. And thus did Taylor's theory of scientific management migrate and transmute itself into the Information Age. The tractor has become jet propelled.

We suggest a litmus test. If you are employing technology to increase knowledge content, allow local empowered action, and remove central control, then you are moving with the currents of the Information Age. Conversely, if you use technology to remove knowledge content and tighten control, then you are extending the Machine Age.

In the West, most are doing the latter. These find that apply new technology to Old Think, (e.g., reengineering) gives temporary relief. However grim and bleak the results, as long as managers can squeeze profits from the old methods they will continue. Still, we think, because of the trends in Knowledge Based Business (Chapter 2), Global Competition (Chapter 3), and Technology (this chapter), all these efforts are doomed to fail. Ironically, in the end, the cross these managers will be nailed to will be of their own making, one labeled "profits." *In an age of pervasive, global, technology-intensive, knowledge-based business, anything that is standard and routine will inexorably fall to razor thin margins and third world wages.*

Fortunately, the free market allows for competition using different approaches. This creates great opportunities for Information Age firms who can offer unique, compelling, and affordable products. Consider how customers rejoiced when technology offered a better alternative, direct broadcast satellite (DBS), to their local cable monopolies, or private sector alternatives (FEDEX, UPS) to using the post office. Price is, after all, only one component of a product. Buying a Japanese car or an Intel based computer is usually a happy experience, though there are always less costly alternatives. When the post office offered cheaper service and money back guarantees, few in business cared. They didn't want their money back, they wanted to know their packages would be delivered when promised.

We don't recommend the practice of "putting jet engines on tractors," but it is being done. In the case of reengineering, it is being done almost universally. The reason that such approaches are appealing are that they offer Machine Age firms a way to improve results without having to give up Old Think or change their basic approaches. That means that these methods will continue. Knowing that most businesses will focus on "cheaper faster" gives Information Age firms a solid touchstone for planning. Some of these will prosper.

Technology Theft

The final way that Machine Age firms can compete with their Information Age rivals is simply to let them invent better technology, and then steal it. That is what the Patent Wars are about. The Japanese, in 1993, deemed our patent system "unacceptable" because it limits their access to U.S. technology. In 1994, they signed letters of agreement with our Commerce Department to essentially convert our patent system into a replica of theirs.[20]

The U.S. patent system, unique in the world and written into our Constitution, was the first time in the history of civilization that "property of the mind" was recognized by law. Our system exists to ensure that the inventor can have a limited monopoly to his inventions so that he may reap commercial benefit. The Japanese system, conversely, has a purpose of ensuring that technology is diffused around their nation for the benefit of their large *keritesu* firms.[21]

Box 4-3

> **"America's patent system is one that promotes technological advancement and individual reward — a winning combination. But our system is under attack. This attack is inspired by those who want to steal American technology."**
>
> Congressman Dana Rohrabacher, 1997

Some are noting that "Where nations once fought wars over trade routes and raw materials, the conflicts in the future will be fought over intellectual property." We agree, but a detailed discussion of the Patent Wars is beyond the scope of this book. Suffice it to say that our government is too ignorant of this important infrastructure issue.[22]

In addition to the issue of legalized theft by patent system "harmonization," as supporters of the Patent Sell-Out term it, there is also the issue of industrial espionage, which is much more widespread than most Western executives realize.[23] In some cases, there is evidence that foreign nations have converted their Cold War espionage agencies to the collection of industrial assets. Many U.S. technology firms have experienced incidents where foreign interests have stolen their intellectual property, and in some cases even chips, for commercial gain.[24] Software piracy is a major problem, for example, with China.

Most Pacific Rim nations (Korea, Singapore, etc.) are very careful to ensure that technology rights are a part of commercial transactions and trade agreements. For example, to sell aircraft in China, Boeing must transfer its technology to their firms.[25] Still, Washington, with its "free trade" agenda, chooses not to dwell on industrial espionage and intellectual property loss or theft. The massive and growing high tech trade imbalances ($32 billion with Japan, $7 billion with China, $8.5 billion with Singapore, etc.[26]) are topics avoided by Washington and the Press (a good fraction of which, we note, is now foreign owned).[27]

Real Progress Takes New Approaches

We've examined the limited options for Machine Age firms that want to improve, but without substantive change and while holding on to Old Think. Firms can cut costs, buy rivals, overlay new technology on their old methods, and they can steal knowledge assets, perhaps legally. These tricks have all been exploited, and, like the diet fads that come and go, they can produce some relief of the symptoms. There are hundreds of management "fix it" books that will further teach and develop these approaches, but this is not one.

The real trick to sustained prosperity is to exploit the new technologies to ensure a continuous flow of compelling products. That is difficult. We have noted that the traditional Machine Age methods for technology sourcing, pipeline and incremental development, are, by themselves, no longer effective. That means that firms must do more, a lot more.

That's hard. "Doing more" is the difference between a diet pill and a lifestyle change based health program. This is more like a 12-step program than a corporate strategy. To compete in the Information Age, firms must transform themselves from Machine Age firms into Information Age firms. The first step is to get past denial. CEOs, and probably about three quarters of the managers in a firm must first develop conviction that they absolutely must change.

Box 4-5

"Do not prepare for change because something has happened; change before it happens."

Jan Carlzon, CEO, Scandinavian Airlines System

Until executives are totally convinced that they must change, most won't. Typically, individuals have to hit bottom in their personal lives before they move past denial. That option is probably not open for corporations. If companies wait too long, they will lack the resources to pull off the change, and will probably go bankrupt or be acquired first. We can't make the first step for you, but we can help once you have chosen this path. The rest of the chapter assumes that you are willing to change and in a mighty way.

Breaking "The Logic of Failure"

We have previously referenced Peter Senge's work at MIT on learning organizations.[28] In general, Dr. Senge has concluded that the only hope for real competitiveness is to have corporations learn as organizations. His preferred methods for learning involve interactive computer simulation of the "real world."

Interestingly, on the other side of the planet, Dr. Dietrich Dorner, a cognitive psychologist, won Germany's highest science prize, the Leibniz Award, for work exploring why trained managers and professionals so often achieve disastrous results, even catastrophe, while doing everything "right." His work has been available in Germany for a few years, but is just now becoming available in English.[29]

Why do trains crash when all the signals are working? Why does a nuclear reactor melt down with a superb team of operators alert and at their posts? Why do planners who "help" third world countries by developing health programs not realize that increased life expectancy and population growth will inadvertently cause widespread death by famine? Why did those who brought DDT to widespread use as a blessing to cure world hunger not realize its horrible side effects on the environment?

Dorner has, like Senge, used interactive computer simulations to explore human behavior, as sincere, intelligent, motivated people try to solve simplified problems of the types common to the world we live in. Unlike Senge, his focus has been not on organizations, but on how decision makers think. Dorner's experimental results, all too often, include the same bizarre actions and catastrophic outcomes that occur in the real world. Often, well intentioned interventions with limitless resources applied left the original problems worse, sometimes catastrophically worse, than if nothing at all had been done.

Surprisingly, Dorner finds the answer is not in negligence or carelessness, but in what he calls "the logic of failure." The root cause of failure, he says, gets down to certain patterns of thought — close to what we call Old Think — such as taking one thing at a time, cause and effect, linear thinking, etc. These were appropriate to an older, simpler world, but they prove disastrous for the complex world that we live in now. "We can't do one thing at a time, because everything has multiple outcomes. We can't think in isolated cause and effect terms, because all situations have side effects and long-term repercussions."

Where we speak of clinging to "Old Think" and reluctance to make fundamental change, Dorner uses the word methodism. "Faced with a new problem, the answer is often simple-minded: we'll do what we have done before," he says. Indeed that is what the Machine Age was all about, repetition and optimization.

Dorner notes the appeal of simple labels. "A simple label can't make the complex nature of a problem go away, but it can obscure complexity that we lose sight of it. And that, of course, we find a great relief." Indeed, traditional MBAs' tend towards bullet charts and one liners, what Senge calls "single-loop learning." Dorner notes that the more information you have, the harder it is to make quick decisions, and says that, therefore, all over the world, organizations have universally tended to institutionalize the separation of decision making from information collection. "The point of all this separation (functionalism) may well be to provide decision makers with only the bare outlines of all the available information ..."

"Anyone who has a lot of information, thinks a lot, and by thinking increases his understanding of a situation, will have not less but more trouble coming to a clear decision. To the ignorant, the world looks simple," said Dorner.

Information Age business, of course, perfectly fits Dorner's problem class. The systems that matter all involve complexity, intransparance, internal dynamics, and are only partially or perhaps even incorrectly understood. Complexity means that all the many elements of the system are interlinked, sometimes with non-linear couplings. Intransparance means that some aspects of the system are hidden from your view. Internal dynamics means that the system runs on rules beyond your control, regardless of what you do, and even if you do nothing. Not coincidentally, the Engines of Prosperity described in the next chapter, are coupled and all exploit strongly non-linear, usually exponential, system behavior.

Dorner has a phrase, ballistic behavior, where managers faced with unclear situations, to prop up their own illusion of competence, tend to fire off decisions, like salvos, without giving much thought to where the cannon balls might land. (Indeed, we have tacitly acknowledged this in U.S. business literature, and some deem it desirable: "ready, fire, aim." Of course, we don't always admit to omitting the last step, but we often do, and some have even explicitly suggested that we should.)

Dorner notes five characteristics of the logic of failure. Managers act without prior analysis of the situation. (Don't many Western managers do exactly that? Is not demonstrating one's crisis management skills a good road to promotion in many firms?) Managers fail to anticipate side effects and long term repercussions. (How many in business or government have pondered the repercussions of downsizing?) Managers assume that the absence of immediately obvious negative effects means that they have taken the correct action. (IBM makes its numbers by having a layoff and pulling next year's mainframe orders into this year.) His last two are that managers tend to get overinvolved with pet projects (Sculley with Newton), and finally, that they were prone to cynical reactions (those laid off deserved it).

The good news is that Dorner says we can change, if we choose to invest the work and effort to do so. While he rejects the notion that there is some secret mental trick that will at one stroke allow the human mind to solve complex problems, we can adapt to the new age if we bother to train ourselves, practice, and get expert help. "Our brains are not fundamentally flawed; we have simply developed bad habits. When we fail to solve a problem, we make a small mistake here, a small mistake there, and these mistakes add up."

Still, in the end, the trick of Information Age business innovation is not reading about it, or training for it, it is actually doing it successfully and consistently. As Carl von Clausewitz, the strategist, said centuries ago, "In war everything is simple, but it's the simple things that are difficult." Compared to his simpler age, the "fog" surrounding what we perceive and the "frictions" interfering with what we attempt are far greater.

Anyway, the point of this section to remember is: Exploiting Information Age technology in business, at least to any great effectiveness, requires very different thought patterns from those of the Machine Age. Until one starts practicing system thinking, it will be difficult to fully develop or exploit the new Information Age technologies.

A New Type of Time

This is a time of fast cycles, and the fetish of the moment seems to be "cheaper, faster." In fact, one of the great appeals of reengineering may be that it offers the promise of not just saving money but of simultaneously cutting costs, laying off people, and running the firm's business processes quicker. "We are not firing people, we are scientifically reengineering our processes to become quicker and more efficient."

So most managers know it is important to go fast, and we agree, but do they know why, when, and how fast? Well, most do at an elemental level, "Now. Real fast. Because if we don't, our competition will kill us." Still, there has always been competition, and only recently has speed been such a competitive advantage for manufactured goods.

The reason that time is suddenly so important is that the key technologies of the Information Age are all non-linear, in fact most are exponential. We think that to correctly exploit these, managers need to better understand time.

We live in four, not three, dimensions, but can only perceive configurations in three of these. We have rich experience in perceiving spatial dimensions, and can look at things, detect patterns and quickly perceive reality. For example, when pouring, we can easily detect when to slow down to prevent overfilling a coffee cup. We have rich experience in how to detect the fullness of a parking lot, or of what parts are missing from a pattern. We can't do this for time.

Successful executives in firms are very aware of the power of geometric laws. Intel, for example, bases its business on Moore's Law that says microprocessor performance will double every 18 months. Andy Grove, not long ago at a Silicon Valley event, told the tale of the inventor of chess in ancient China. (Dorner uses the same example, but he says it was India. No matter.)

The Emperor condescendingly offers a reward to the inventor for his great feat. The inventor can have any item he wants from the royal treasury. The inventor, annoyed at this patronizing reception of his great achievement, replies that all he wants is a little rice. One grain for the first square on the chessboard, two for the second, four for the third, and so on for all the squares of the board. The Emperor, delighted to get off so cheap and laughing up his sleeve at the stupidity of the inventor, agrees to give the reward.

Grove said that there were two endings to the fable. One was that the Emperor had the inventor killed, and the other was that he kept his pledge and the inventor took over the kingdom. You are encouraged to run the math yourselves. The last square alone would have 2^{63} grains of rice. That is about 153 billion tons. Put differently, it would take 31 million cargo ships full of rice, if each ship held 5,000 tons. The power of exponential trends is awesome.

Let's use what we've learned about exponential growth to explore possible futures. One exponential trend is growth in the U.S. trade imbalance. Another is compound interest. Greider reports a crossover point in 1994 for the United States. In that year, for the first time since 1917, our nation's total payments to foreign investors (stock dividends, bond payments, royalties, etc.) exceeded those to domestic investors.[30]

Box 4-6

"What's the most powerful force in nature? Compound interest." Albert Einstein

The numbers are, like the first grains on the chess board, small. Still, barring a major change in national trade policy (which no one presently anticipates) the trend will continue. Greider says this small event signals that the U.S. in now like a family living off its credit cards, paying off each month's bill by charging it and going deeper in debt.

Naturally, as a nation we have many more assets and pay a lower interest rate than a family. Still, the end result is ruled by the same simple, inexorable, exponential mathematics. At some point in the not very distant future all our nation's credit will be used up, the balance will keep growing, and we will tap out. So what?

One scenario is that new U.S. leaders emerge in time, educate the public about the threats and opportunities presented by geometric trends, and mount major initiatives to fix our societal problems, regain our competitiveness at building innovative, compelling products, and thus redeem our debt. That could represent a new golden age for our society.

Alternately, Greider quotes economists who think Japan's strategy is to keep the U.S. from doing anything drastic for ten or fifteen years, "until the Asian economies develop and they no longer need our market." That scenario, collapse, war, or economic conquest, is darker.

This would be a good research topic, and one that is amenable to the types of computer simulations that Senge and Dorner have been doing. There may be other possible outcomes, including some beyond our present ability to imagine.

Box 4-6

"Despite the second law of thermodynamics, decline is not incvitable. The new science of dissipative structures proves it. Life eats entropy!" Barbara Marx Hubbard, Futurist, Wilsonville OR 1997

For example, consider other nonlinear societal trends. Recall Grove's chessboard talk? His point was the fact that the knowledge of the human race is doubling at a geometric rate. He concluded that humankind was now at the 32nd square of the chessboard. (Actually that understates the case, because, unlike the chess board, the doubling time of human knowledge is <u>also</u> decreasing at a geometric rate. The doubling time of knowledge has gone from 1,500 years at the birth of Christ, to about 6 years today. It trends to zero at around 2012.[31])

Other hopeful trends go beyond exponential to quantum. Paul Ray of the Institute for Noetic Science says the fastest growing subculture in the U.S. is "cultural creatives" who now number over 44 million people, almost 25% of the voting population. These people are convinced the old Megastate and Machine Age bureaucracies are dysfunctional and are eager to move to the next stage of human evolution.[32]

Technology Attitudes. What Matters?

In a world seemingly swimming in technology, including myths and rumors of technology, what should a firm do? Where should R&D funds be invested? How should R&D be managed?

The first helpful realization is that most of what you know about R&D, including virtually everything that used to work, is no longer valid. Here is where you must work the hardest to reject Old Think. The new center of business is knowledge creation, and that takes R&D.

Box 4-7

> **"For a successful technology, reality must take precedence over public relations, for nature cannot be fooled."**
>
> Richard C. Feynman, Nobel laureate.

Bill Gates, who had only a few years before, accused anyone touting the Internet of being Communist, because it made no sense to give products away free, woke up and changed. He turned his whole firm on a dime, and the new Microsoft strategy is to "embrace and extend" to encompass the Internet. That is a top level strategic decision, not an operational decision, and it is, at the core, a technology development decision. In the same vein, Andy Grove bet Intel on his decision to abandon the larger and growing DRAM market and move to microprocessors, an exponential trend he could control. That was a strategic, not an operational, decision.

Conclusion: Just as war is too important to be left to the Generals, technology is too important to be left to the technologists. The key technology investment decisions are strategic, and are, or should be, tightly integrated with a firm's other strategic decisions.

The next part of making technology decisions is that they, again like war, are not things where general rules or statistical 1averages are of much help. To quote Clausewitz again, "War, in its highest forms, is not an infinite mass of minor events, analogous despite their diversities, which can be controlled with greater or lesser effectiveness depending on the methods applied. War consists rather of single, great, decisive actions, each of which needs to be handled individually."

This gets us directly to another part of Old Think to cast off. Technology decisions are not repeatable. The idea of using Machine Age process to make technology decisions is nonsense. What happened last time, and what was done by this predecessor, or that famous executive is useful knowledge that might or might not apply to the case at hand.

Conclusion: You only get to create a new technology once, and the next one is different. Learning how, say, Bill Gates did Windows, is not particularly helpful in telling someone how to do the next great technology for the computer industry.

Finally, technology decisions are, not only not repeatable, they are also unique. The Information Age is about creating new and prosperous futures. A business wins over the competition only because it does not repeat a past approach or follow the same path its competitors are pursuing.

Conclusion: There are no golden rules and models for technology decisions. Or, rather, there are many good rules and models, but knowing when to apply them makes the difference between success and failure.

Right now Japanese high tech firms, on average, spend 80% more on R&D than they do on capital equipment. Japan's National Institute of Science and Technology Policy (NISTEP; we have no equivalent and do not collect such data) produced a 1991 survey of their industrial base, Kodama himself was a co-author.[33] 80% of the fabrication industry firms surveyed thought the ratio of R&D to capital investment expenditures would stay the same or increase (40% said "increase"), and the remaining 20% "didn't know."

Kodama's book proposes six specific dimensions of paradigm shift as "sources of Japan's technological edge." Manufacturing should shift from producing to knowledge creation. Business diversification should shift from spin-off to trickle-up. R&D from dominant design to interindustry competition. Product development from pipeline to demand articulation. Innovation patterns from breakthrough to technology fusion. Finally, societal diffusion should shift from technical evolution to institutional coevolution.[34]

Technology fusion is almost the reverse of Western breakthrough invention. The concept is to take (critics say "steal") existing technologies from outside sources, and blend them together for unique competitive advantage. There are Western examples of this practice. Bill Gates bought DOS for a pittance, "borrowed" the Windows interface from Xerox, and NT from DEC. Still, except for mergers and acquisitions, most Western firms are internally focused.

Western firms are organized for and focused on "pipeline" product development. The same basic products evolve by model year. Detroit's cars, even now, change little from year to year. Intel's microprocessors evolve very rapidly, but still in a pipeline. Demand articulation is looking below the surface to create totally new product types. That yields CD players, home video recorders, Walkmans, etc. Large Western firms, at best, do that sporadically. Kodama claims that Japan has developed this into a national system and extended it into public policy.

Spin-off versus trickle-up is another strategic dimension of choice. The West, seeking quick pay-back, stereotypically picks the most complicated and expensive applications for a new technology. Japan prefers low-cost mass markets. Hence, where the West targeted carbon fibers for military aircraft, the Japanese started with sporting goods. Japan has won market dominance in carbon fibers, flat panels, and several other large, fast-growing markets that the West invented.[35]

Knowledge creation is perhaps the Kodama paradigm most hopeful for the West. Canon needed breakthrough cost reductions for its copiers. Instead of Western reengineering, downsizing, and squeezing more from less, they committed top talent, invested deep thought, spent patient money, and looked all over the world for technologies that would solve their problem.

The copier drum, the source of 90% of copier maintenance problems, had to be precisely round and so very cheap that it was disposable. The breakthrough came at a picnic, when one worker held up his beer can, exclaiming the Japanese version of "eureka." Sure enough, adapting those processes and technologies led to major advantage. Canon print engines now power virtually all the copiers and laser printers in the world.

If we wish to control the markets of the future, those built from knowledge, those where the Engines of Prosperity reign, we are starting late. The sleeping giant is, at best, just starting to stir. R&D investments in the U.S. have been declining, especially for basic research. Few companies are calling Congress to defend the patent system.

We did some work with colleagues in Price Waterhouse, and were surprised to discover how little CEOs knew about the effectiveness of their R&D investments. This led to a 1997 international conference "hot topic" panel on metrics for innovation.[36] (Conversely, the typical CEO knew his firm's stock price, financial performance, and operational metrics precisely and intimately.)

What Is Technology (Viewpoints)?

The root word is the Greek *teknologia,* meaning the systematic treatment of an art or a craft. Hence, the methods for managing Information Age business are, of themselves, technologies. This book could be thought of as an early book about a New Science, the technology of Information Age, as opposed to Machine Age, management.

The common Machine Age definition of technology is much more narrow. "The application of a science, especially to industrial or commercial objectives." This definition leads one to thinking of tools and the processes and science from which we build products. That is part of it, as well, but a diminishing part. As we have discussed, overlaying Information Age technology on the old structures and models only helps to a point.

At the highest level, technology shapes civilization — how societies are structured and how people think and interact with each other and the world. Just as tribal societies and then feudalism were successful Agrarian societies, and Megastates were successful Machine Age societies, more advanced societal forms will undoubtedly evolve for the Information Age. A new golden age is possible, but staying the same is not. Mostly likely, as with past transitions, people, organizations and nations will either embrace the new technologies and move forward or they will cling to the old and regress.

Box 4-8

"Persons grouped around a fire or candle for warmth or light are less able to pursue independent thoughts, or even tasks, than people supplied with electric light. In the same way, the social and educational patterns latent in automation are those of self-employment and artistic autonomy."

Marshall McLuhan (1911–80), Futurist

What Technologies Matter?

At least some things have become simpler: You should look for technologies that offer mass markets and very steep learning curves, regardless of the front end investment needed. Why? Because, your competitive advantage through knowledge comes through exploiting these geometrically increasing economic returns.

Box 4-9

> **There are two types of technology: exponentially developing technology and unimportant technology."**
>
> Ian Rose, when head of the old Bell Labs

The list is fairly obvious, and Japan's MITI, inevitably, did a white paper on the subject. They deemed the key technologies to be microelectronics, software, materials science, machine tools, aircraft, biotech, etc. Basically, in the Information Age, you look for technologies that allow products to be made from knowledge, for these have the very steep learning curves. These allow high profit mass-market products.

The other technology sets are those that let you manage and operate a globally dispersed Information Age business more effectively. These range from PCs and cellular phones, to intelligent databases, to, perhaps, things like video conferencing. A good example of a prosaic business reinvigorated by new technology is <www.amazon.com>, the electronic book store. The trick is to use the new technologies to empower creativity instead of imprisoning workers into routine. That is harder and often missed or ignored. The new interactive computer simulation technologies, per Senge and Dorner, which facilitate new thought patterns and organizational learning can be of much use, both for training and for strategic contingency planning.

The new type of R&D is much broader. Wealth does not come from technology creation, but from technology application. Whether you accept the Japanese models or not, effective R&D clearly includes a crucial business role — what we call the "guru" and what the Japanese call "demand articulation." To make a lot of money, you must discover latent needs before even the customers are aware of them, and that is a very difficult task. It is usually the most difficult task. Market creation is as important as technology creation (e.g., VCRs, FAX machines, cellular phones, CDs).[37]

In contrast to the Machine Age, you have to look outside your firm. Bill Gates did not invent DOS or Windows or NT, he acquired them from external sources. The Japanese did not invent transistors, they licensed them from Bell Labs. Intel got the idea for the microprocessor from a Japanese source, and so it goes. Most successful firms already involve "outsiders" in their new product development, and this trend will increase.

That is good news for the smaller and weaker firms, who lack the resource of the multinationals. For these with limited resources, the strategy is to latch on to an exponential wave and surf it. Instead of building your own computers, use PCs. Instead of building your own proprietary systems, use the Internet.

A Scorecard for Innovation

There are, as yet, no generally agreed to and useful metrics for Information Age innovation or technology development. Still, some type of score card should be presented. We have chosen to (subjectively, but based on anecdotal experiences with hundreds of firms) use the following eight criteria to rank the hypothetical average U.S. firm:

1. Infrastructure. (Educational, freedom from crime, communications, government support, etc.)
2. Intellectual property (IP) Law. (Ability to protect intellectual property assets from larger predators.)
3. Technology Base. (Are trained, seasoned engineers and scientists available?)

4. Ability to fund R&D. (Can you invest in R&D without a return for several years?)
5. Operations. (Cycle time, efficiency, production capacity)
6. Desire. (At a corporate level, and with high conviction, is innovation truly your top corporate priority?)
7. Culture. (Do individuals in your society center on wealth creation through innovation?)
8. Teams. (Do you utilize autonomous, empowered, multi-disciplined teams?)

We have rated these criteria on a scale of 1 to 5 (if it were letter grades, 5 = A, 1 = F). The average global competitor would, of course get a "C" or a "3" rating. Figure 4.1 gives our impression of how the average U.S. firm is doing versus its global competitors. Since some of these ratings may seem extreme, we should explain. Though these "grades" are subjective, they are reasoned and as objective as we know how to make them.

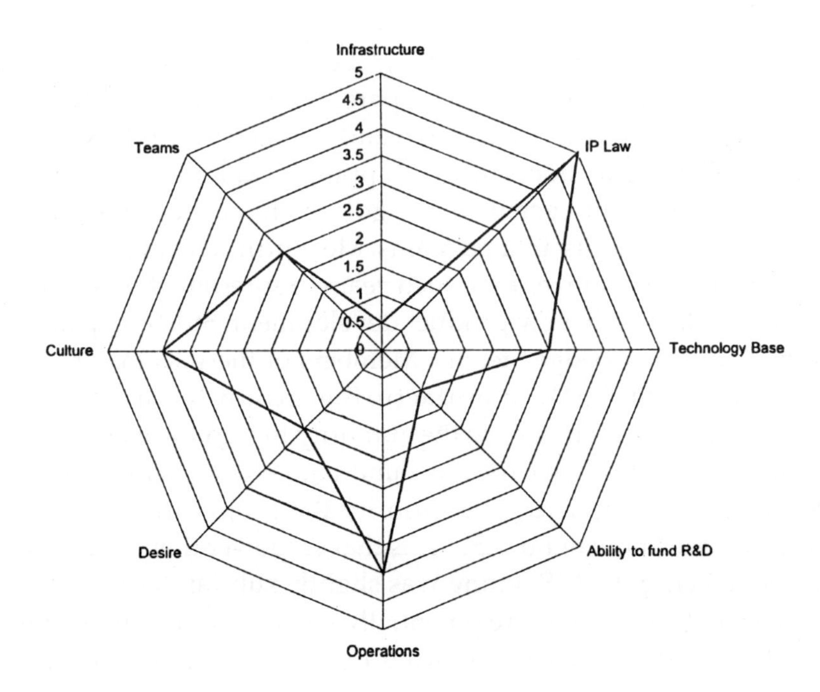

Like the era when the U.S. was losing major markets because of cost and quality, the best U.S. firms are as good or better than any in the world. The problem was that the average firm was sub-par and uncompetitive. Much the same situation exists today for U.S. Information Age technology development capability.

Firms like Intel, Hewlett Packard, and Motorola locate in safe, low-crime areas, their buildings are secure, and they provide their own high bandwidth communications links and multi-level employee training programs. They cast out societal misfits, and focus on getting the best people, and bringing out the best in the people they have. The best of these firms would, of course, get a 5 in all dimensions. Smaller, less prosperous firms cannot afford these things, and must depend on local colleges, phone companies, and public infrastructure.

Also, some large firms may choose to view issues like infrastructure support for competitive advantage as outside their business mission. Or they may take viewpoints that make this subordinate to other objectives. For example, colleagues have assured us that at least some IBM facilities discourage discrimination against the incompetent as a part of the formal "diversity training" being provided to workers and professionals. Also, when we compare, we must include places like Singapore (near zero violent crime and vandalism, 100% wired with fiber, excellent education, superb shipping) and Japan.

The U.S. has in recent years gotten much better at fielding empowered, multi-disciplined teams. Most firms recognize, or at least say that they recognize, that such teams represent the best known method for fast cycle new product development. Still, companies like Sharp have "gold badge" teams. Members of these teams are chosen from corporate R&D, manufacturing group laboratories, and engineering. One member has the authority to lead the team, and each member of the team wears the same badge color as the company president, gold, signifying the priority the project has on corporate resources. We know of no U.S. firms that empower their teams to such an extent, hence the U.S. rating was slightly substandard.

Some of these ratings are potentially volatile. For example, the U.S. patent system has, by far, produced the best record of innovation in the history of the world.[38] That is as good as it gets, so that was rated as a "5," thus raising two questions. First, should intellectual property (IP) protection be included as a integral part of infrastructure, where the U.S. rated very low? Second, is that an advantage that U.S. firms can count on?

Our reason for including IP protection as a separate category is that if the only competitive advantage in the Information Age is knowledge (technology), then the ability to protect this key asset from piracy surely must have exceptional importance. Hence, we included it. The unique U.S. patent system may represent the greatest national competitive advantage we have. However, whether or not we can count on it depends on what happens in Congress. The advantage could be gone by the time this book is published. Each year since 1994 there have been bills introduced to "harmonize" our patent system and bring it down to international norms. At present, bills to do so (the 1997–98 versions are HR 400 and S. 507) are being argued in Congress. Much as managers hate to admit it, what government does or does not do greatly impacts business.

Finally, we note that some dimensions of innovative advantage, however real, are notoriously hard for large firms to exploit. The United States has always been a hotbed of individual innovation. No other nation has anything like our Venture Capital community or Silicon Valley. Still, the Machine Age firms have always tended to suppress innovation. As a result, many large firms are ringed with smaller companies (and some larger) whose founders left to develop technologies their former employers rejected or ignored.

PART II
EXAMPLES, RIVALS, AND PHILOSOPHIES

Chapter 5

The Engines of Prosperity: Foundations

Key Themes at a Glance

In the first part of this book, we challenged conventional assumptions and methods of management. "Old Think," defined as deep seated assumptions and beliefs on how to succeed in business, have become poor guides to action in today's world. In business corporations, these include the primacy of "top-down" thinking, the undue emphasis on the bottom line, the belief in sustainable competitive advantages, and "Band-Aid" solutions to management practices.

New thinking is needed to displace Old Think. The dawning of global markets, the pervasiveness of technology, an enlightened work force, and new competitors are driving pressures to change.

In this chapter, we discuss traditional and conventional approaches to management, rooted in classical economics and competitive strategy. These include the military origins of strategy, classical rationalistic viewpoints, gap analysis, portfolio theory, and industry analysis.

We also provide criticisms of the rational school, notably advanced by Mintzberg, and discuss incrementalist approaches to strategy. We raise issues about the adequacy of these models in

regard to different cultural contexts and knowledge-based industries.

We present leading-edge theories that include core competency, strategic intent, and hypercompetition.

In discussing hypercompetition, we present original arguments advanced by D'Aveni, and how they provide a point of departure for developing the Engines of Prosperity.

We then present the foundations of a new competitive environment — one for which *Engines of Prosperity*, or recurring patterns of growth and renewal, become the viable alternative to Old Think. Understanding the dynamics of these Engines, and learning to compete within them, form the imperatives for the new competitive order.

Introduction

In Lewis Carroll's *Alice in Wonderland*, Dorothy — the story's leading character — seeks advise on directions from the Queen of Hearts. When asked where she is going, she replies that she did not know. Then, the Queen of Hearts declares: "If you don't know where you are going, any path will take you there."

Strategic planning addresses the competitive strategies of corporate firms. It is a field that attempts to understand and diagnose changes in a firm's external environment, and how a firm might then position itself against such changes in unique ways for it to maximize on its strengths and to minimize its weaknesses. In fundamental ways, strategies define what a firm is all about. Strategies provide consistency and cohesion to a firm's actions and behaviors. Strategies also defines what makes a particular firm different or unique from others. In short, strategies provide the answer to Dorothy's quandary: they tell you where you should go and how to get there.

While such descriptions of the field of strategy are soothing in tone and reassuring in content, traditional and conventional theories of strategy are in a critical policy juncture. In our last three chapters, we have provided emerging challenges for the 21st century manager: the advent of a knowledge-based society, the pervasiveness of new technologies, and the dawning of globalization. These have become imperatives for significantly new ways of managing enterprises.

Traditional theories of strategy have be used to explain the growth and dominance of market leaders, such as IBM, DEC, General Motors and Sears Roebuck. For the most part, these firms built their advantage on the basis of economies of scale and scope. Once considered unassailable, however, these firms have given ground to Microsoft, Toyota, and Wal-Marts. As such, the weathered strategy rule book on how to manage — one premised on a firm's ability to sustain its competitive edge — is now challenged by new theories of strategy that allege that competitive advantages can no longer be sustained, at least for a long time. The traditional theories fall short in describing the "new rules" of the new competitive environment, and how to compete in it. In this chapter, we describe the evolution of strategic thought, and we argue why traditional and conventional templates for managing in this environment are limited in a competitive environment where knowledge and technology pace the development of new industries. This discussion provides the basis for *Engines of Prosperity*, or recurring patterns of growth and renewal, that are formally developed in the next chapter.

The Initial Promise and Lure of Corporate Strategy

Box 5-1

> **Strategy is the great work of the organization. In Situations of life and death, it is the Tao of survival or extinction. Its study cannot be neglected.**
>
> Sun Tzu, *The Art of War*

In much of the Western world, the concept of "corporate strategy" has been the imperative for organizing competitive order. In a 1859 classic, *On the Origin of Species*, Charles Darwin stated: "Some make the deep-seated error of considering the physical conditions of a country as the most important for its inhabitants; whereas it cannot, I think, be disputed that the nature of the other inhabitants with which each has to compete is generally a far more important element of success."[1] This statement provides a fruitful starting point for the study of business strategy.

Military Origins

Despite its common use, strategy is a commonly used word and is often misunderstood. *Webster's New Dictionary* defines it as "the science of planning and directing large scale operations, specifically of maneuvering forces into the most advantageous position prior to actual engagement." As intimated above, the term "strategy" originates in the military, specifically from the Greek-word, *strategos*, meaning "general."[2] As wars became costlier and more complex, it became obvious that victory depended on factors other than size, bravery, and prowess. Throughout history, there is compelling evidence that the outcomes of individual battles, military campaigns, and business success often depended on the highly subjective evaluations of people's intentions, capabilities, and behavior.

Liddell Hart, who is a scholar in this area, states that "the true aim of strategy is not to battle but to achieve such an advantageous position that, if it does not of itself bring the enemy to surrender, would produce a sure victory in the battlefield." He condenses military strategy to a single word: *concentration* ("concentration of strength against weaknesses").[3]

The rationale for competitive business strategy is rooted in two interrelated concepts: *competition* and *uncertainty*. Competition is directed at understanding industry conditions and one's competitors. Without competition, it's difficult to talk about strategy (although it still makes sense to plan). Competition is also rooted in the lack of resources for the parties involved. Uncertainty is the driver of change, and the reason for most plans. One anonymous writer noted: "The future isn't what it used to be." Or Kittering once said: "I'm interested in the future because I will be spending the rest of life there."

Educational Origins

As early as 1911, Arch W. Shaw, a Chicago-based publisher whose flagship publication was *System* — " the magazine for management" (later purchased by McGraw Hill and renamed *Business Week*) — complained to Harvard Business School Dean, Edwin Gay, that the School's courses were too general and descriptive.[4] Its students "wouldn't recognize a problem if they saw one." Both agreed that the concept of a required second-year course, to be called Business Policy be initiated within the School. Shaw adopted the vantage point of the upper-level manager in getting the students to "recognize the problem." He secured the cooperation of fifteen senior managers, each of whom agreed to present a real problem to the class over the course of a week. In the first of three sessions, the businessman would explain his problem, and answer students' questions regarding it. In the second class meeting, two days later, students would hand in a written "problem analysis," with a recommended solution. In the third session, at the end of the week, the businessman would discuss the reports.

The Influence of Management Philosophers

In the course of American business history, various management philosophers, such as Frederick Taylor, Henry Fayol, Elton Mayo, Chester Bernard, and Peter Drucker shaped this dynamic that was business policy into a practitioner-oriented, long-term planning method.[5] As inflation, high interest rates, recessions, and technological innovations continued to impact corporations, long-term planning became necessary for them to survive, remain viable, and to prosper. Strategic planning evolved from problem recognition (as initially envisioned by Arch Shaw and Edwin Gay) to systematic, formalized planning: it took form of a comprehensive statement, expressed or implied by management, of the long-term objectives and purposes of the organization, an assessment of competitive factors that may influence these objectives, and current plans that detail the utilization of resources to achieve the organization's objectives.

Box 5-2

> **The determination of the long run goals and objectives of an enterprise, and the adoption of courses of action and the allocation of resources necessary for carrying out these goals.**
>
> Alfred D. Chandler, Jr., *Strategy and Structure*

Previous treatises on strategy extolled the virtues of discipline and sustaining competitive advantages. While not subscribing completely to the tenets of "rational economic man," these works view managers as deliberately rational, if not purposeful decision-makers. Firms are seen as acting in ways to maximize profits, and that relatively efficient markets emerge over time.

In this context, strategic choices matter.[6] Good planning is what it takes to master internal and external environments. Rational analysis and objective decisions make a key difference in determining the long term success or failure of the firm. Organizations are seen as monolithic entities, led by able and visionary executives. Accordingly, formulating strategic choices form the critical task. Strategic implementation is conceived as co-aligning strategic opportunities and requirements with appropriate structures, processes, and even corporate cultures. This is not seen as problematic.

Strategic leadership is vested in a visionary top manager who formulates strategy and oversees its implementation. It subscribes to the omniscience of Ken Olson, CEO of Digital Equipment, who once said: "My job is to make certain that we have a strategy and that everyone understands it." It celebrates the heroism of Percy Barnavik (CEO of Asea, Brown and Boveri) who would carry 2,000 overhead transparencies to communicate his vision to his company, earning him the reputation as the "overhead president."

The Traditional View of Strategy Formulation and Implementation

Traditional planning typically encompasses three critical phases: (1) Identifying opportunities and threats in the external environment faced by an organization; (2) Matching these elements against the internal capabilities of the organization, and developing key strategies; and (3) Developing appropriate structures and processes to enhance the implementation of these key strategies.[7]

Early Models: SWOT Analysis, Gap Model, and Distinctive Competence

The application of the conventional view is perhaps best represented in the SWOT (Strengths-Weaknesses-Opportunities-Threats) Analysis. The alignment between external environment and internal resources is evaluated on the basis of this analysis. A derivative of this approach is another popular model called the *Gap Model.* Developed by the Stanford Research Institute (SRI), the gap model seeks to define the gap between where the client wants to be and where its current operations and profits are taking the company. The planning process then leads to developing appropriate strategies to close the gap.

An outgrowth of SWOT Analysis and the Gap Model is the *Distinctive Competence Model,* popularized by the Harvard Business School. This approach defines a client's strong point relative to its competitors and anchors key strategies on this strong point (called 'Distinctive competence'). What does the client do well and better than any other in the industry? How might the client utilize this distinctive competence in tandem with opportunities in the external environment to improve its competitiveness?

Some examples of firms with distinctive competencies include General Electric (reputation and service in home appliances), Netscape (Internet navigator systems), Design and Manufacturing (low cost dishwashers), Intel (microprocessors), IBM (dealership and service), and Microsoft (operating systems).

Later Models: Product Portfolio and Experience Curve

These models and analyses became household applications in many corporations, bolstered by management consulting firms that were now heavily populated with newly-minted MBA graduates. In time, models were both refined and embellished. Two models became particularly attractive in the 1970s through the 1980s: *Product Portfolio Planning*, and the *Experience Curve*.[8]

Developed by the Boston Consulting Group (BCG), the product portfolio matrix consists of categorizing a client's products and business according to market share, profitability, and growth potential.[9] Generally, four groups emerge: *stars*, those products with high growth potential and high market share; *cash cows*, those with lower growth potential but high market share; *question marks*, those with expected high growth potential but presently with low market share; and *dogs*, those with low growth potential and low market share.[10]

Typically, the 'stars' and 'question marks' are exploited for the future; 'cash cows' provide the cash for the 'stars' and 'question marks'; and 'dogs' are milked and divested. While the approach is still popular, it has been criticized as unduly simplistic and difficult to implement.

Arthur D. Little (ADL) and the Boston Consulting Group (BCG) popularized the Experience Curve Model that essentially tracks the progress of a product from a high start up cost and low earnings position to an eventual standardization using low-cost production techniques that yield higher profitability.[11] Lower cost is used to establish a formidable competitive position particularly once product maturity is reached and the emphasis shifts to price competition.

Early Criticisms: Restrictiveness

The logic implicit in portfolio planning is for cash cows to become the financiers of other developing products in a firm's portfolio. Ideally, cash cows are used to make question marks into stars. Since doing so requires a great deal of cash to keep up with rapid growth as well as to build market share, the strategic decision is which question marks should be targeted for stars. Once a star, a product becomes a cash cow once its market growth starts to decline. Question marks that are not selected for investment are harvested (managed to generate cash) until these become dogs. Dogs are, of course, harvested and eventually divested from the portfolio. The firm should manage its portfolio in this desirable sequence such that the portfolio is in cash balance.

The applicability of the product market portfolio in any industry has been questioned in terms of the validity of its assumptions, its basic design, and its ease of implementation. Some question the purported link between market share and profitability, asserting that the link may be weakened because of the following reasons: (1) low share competitors (late comers) may also have the steepest experience curve, e.g. Japanese; (2) low share competitors may have low cost suppliers or some in-built cost advantages; (3) the product has been defined properly to account for important shared experiences and other interdependencies with other products, but not all products have experience-related costs; (4) large share firms may be subject to more government regulation; and (5) it is not true that the most fruitful markets to enter are those with the highest rates of growth.

In addition, the technique has also been faulted in terms of its rather simple two-variable formulation — the rate of market growth and the relative market share. It is argued that, while these constitute important factors, they do not necessarily determine a product's success or failure. Similarly, while market growth is a good proxy for required cash investment, profits and cash flow depend on a lot of other things. Recently, the Boston Consulting Group has advanced a reformulated matrix, partly based on these criticisms regarding the simplicity of its initial formulation.[12]

The third area of criticism is the assumption that the portfolio can be easily implemented. It is difficult to unidirectionally take funds from a cash cow for use in question marks, nor is it easy to divest a dog that might have contributed to the firm's fame and success in the past. While concepts like synergy and portfolio appear nice and logical on paper, these may not be well received in a world that is fraught by interdivisional politics, territorial power, and jurisdictional concerns.

Even so, the above criticisms do not suggest that the product market portfolio matrix is not useful. Rather, they suggest that there is ample opportunity to reformulate the basic design which has been done by the Boston Consulting Group. In fact, limitations of the technique might have motivated the development of improved matrix formulations, such as General Electric's Industry Attractiveness Matrix. The criticisms may also suggest that the BCG matrix is useful, but in a more restricted scope. Subsequent studies on the PIMS data base indicate that market share and profitability are strongly linked in *stable* market situations.

Mid-Range Criticisms: The Elusiveness of Strategic Fit

Traditional theories suggest that strategies have to be consistent or demonstrate "fit" with organizational structure and processes for the organization to succeed. The empirical foundations of strategic consistency can be traced to the research by Alfred Chandler in 1972.[13] On the basis of an historical study of seventy of America's largest firms, he proposed that organizational structure is determined by the growth strategy of the firm. That is, as firms move from volume expansion, vertical integration, to diversification, their structural forms evolve from simple functional units to multidivisional firms. Chandler's research suggested that for each type of growth strategy there was an appropriate organizational form. In fact, companies that had a good "fit", or consistency between strategy and structure, tended to financially outperform firms that did not.

The emphasis in these models is developing adequate structures and processes to support the growth strategy of an organization. As an organization evolved over time (i.e. from embryonic to growth and maturity), it needs to adjust its structure, planning systems, rewards, and compensation in accordance with its growth requirements. The objectives of stage-of-growth models are to identify pressures that occur in these transitions, and to systematically modify the appropriate structures and processes to smooth out such transitions.

Metamorphosis models are typified by the approach suggested by Greiner in 1978.[14] Age and the size of organizations provide the driving force that underlie metamorphic change. Therefore, as organizations develop in terms of size and age, they face different types of "crises" which are overcome with more complex structures and control processes. In spite of its intuitive appeal, the model has been criticized as oversimplistic and lacking in specificity.

Galbraith and Nathanson provide the necessary embellishment in their "stage-of growth" model.[15] Based on an extensive review of the experiences of both American and European companies, Galbraith and Nathanson expanded the works of Chandler, Lawrence and Lorsch, and Greiner which takes into account global strategies. The key premise is the same as Chandler's — that is, starting from a simple form, any source of diversity may be added to evolve into a new form. In their formulation, Galbraith and Nathanson see organizational structures evolving to meet the requirements of a changing growth strategy. Internal processes such as performance rewards, measures, careers, leadership styles, and organizational choices are used to support the requirements of strategy and structure.

Congruence or consistency between strategies, structures, and processes is the critical element that underlies this stage-of-growth model. In other words, firms offering multiple products in distinctly different markets would need a multidivisional structure with the corresponding processes. The failure to adopt such a structure would lead to inefficiencies. Similarly, it would be quite costly for firms that offer a few related products in similar markets to have a multidivisional structure. A functional structure with some controls would be more efficient and effective. Overall, these studies have provided the criteria in which to assess linkages between strategy, structure, and processes.

Despite its intuitive appeal, the concept of strategic fit has not held its ground under empirical tests. The landmark studies by Lawrence and Lorsch were criticized for low validity.[16] Subsequent efforts to replicate these initial studies resulted in rather mixed support for the congruence hypothesis, prompting several authors to declare the concept as theoretically inadequate and practically implausible. More recent attempts to restudy the concept using highly advanced statistical methods and techniques have proven to be inconclusive.

The Strategists Respond: Industry Analysis and Competitive Behavior

Box 5-3

Two hikers — the classic strategist and the competitive strategist — encountered a menacing bear. As the bear advanced, the classical strategist started to collect branches, stones, and sharp objects to prepare for mortal combat. The other laced his running shoes. When told he could never outrun the bear, the competitive strategist laughed. "No problem. All I need to do is outrun you."

Well aware of the criticisms directed at strategic planning in the seventies that was derived from particular ideologies ("rational economic man") and heuristics ("strategy in practice"), Michael Porter responded with a systematic framework rooted in microeconomics and competitive behavior (i.e. the Five-Forces Analysis).[17] He explains the failure of many companies as the result of failing to "properly" formulate strategy — a weakness that would presumably be remedied by the application of his new framework. For Porter, therefore, strategy refers to interlocking decisions by top management over time aimed at developing a sustainable competitive advantage. The success of Porter and the spectacular rise of his consulting firm, Monitor, attest to the growing popularity of his ideas, that is, notwithstanding Mintzberg's criticisms.

Michael Porter's authoritative work focuses on industry dynamics as the key element of environmental analysis. The state of competition in an industry (defined as a group of firms producing products that are close substitutes for each other) depends on five competitive forces: threat of entry, intensity of rivalry among existing competitors, pressure from substitute products, bargaining power of buyers and suppliers. Profit potential, measured in terms of long run return on invested capital, is directly related to these forces. Intense forces, such as tire, steel, and paper industries, generally result in low returns. Mild forces, found in oil field equipment and services, cosmetics and toiletries industries, generally experience higher returns.

The strength of each of these forces vary across industries, and it is the task of the strategist to evaluate the critical dimensions of these forces. High entry barriers (pharmaceuticals, office equipment, oil refineries) favor incumbents, even when high profits begin to attract potential entrants. Rivalry among existing competitors takes the familiar form of jockeying for position — using tactics like price competition, advertising, product introductions, and increased customer service to gain advantage.

In most industries, firms are mutually dependent and therefore moves by one firm are often met by countermoves which, if they escalate, may disadvantage the entire industry. Substitute products limit the profitability of an industry by placing a ceiling on the prices which firms can profitably charge. Finally, the bargaining power of buyers and suppliers can create advantages or disadvantages for incumbents, depending on the latter's concentration, importance, and threat of backward or forward integration.

The goal of competitive strategy is to find a position in the industry where the company can best defend itself against these competitive forces or can influence them in its favor. Competition in an industry continually works to drive down the rate of return on invested capital toward the competitive floor, or the "free market rate." The strength of competitive forces determines the degree to which the inflow of investment occurs, drives the return to free market level, and cuts the ability of a firm to sustain above average growth.

There are many different competitive strategies for dealing with the five competitive forces and thereby obtaining a superior return on investment. Firms build advantages in distinct ways.

Box 5-4

> **When astronaut John Glenn was asked how he felt before blasting into space, he replied, "Nervous. Wouldn't you be? I am sitting on top of a rocket consisting of 50,000 parts, each of which was made by the lowest bidder."**

Cost leadership became increasingly popular in the 1970s. It has the advantage of protecting the firms against all five competitive forces since lower prices can only erode profits so that those of the next most efficient competitor are eliminated and the less efficient competitor will suffer first. However, cost leadership requires high relative market share or favorable access to raw materials. It also calls for vigorous pursuit of cost reductions, tight cost and overhead control, and the avoidance of marginal expenses.

Differentiation is the creation of something that is perceived industry-wide as being unique. It does not allow the firm to ignore costs, but rather does not make them the main focus. It enables the firm to manage the five competitive forces by insulating it from competitive rivalry because of brand loyalty by the customer, lowering price sensitivity. The drawback of differentiation is that it requires a perception of exclusivity that is incompatible with high market share.

The third generic strategy is *focus* — on a particular buyer, segment, or geographical market. The drawback is that it necessarily involves a tradeoff between profitability and sales volume and may involve a trade-off with overall cost leadership.

There are two fundamental risks involved in pursuing a generic strategy. The first is that the firm will fail to attain and sustain the strategic position. The second is that the industry infrastructure will change through evolution thereby eliminating the gained advantage.

To redress the deficiencies in examining strategic fit, Porter focuses on *strategic discipline* as what undergrids strategic implementation. For the cost leader, for example, it is essential that it focuses on costs and avoids needless expenses. For the differentiated firms, it is the deep understanding of buyers' needs and expectations. And, for the focused firms, it is the attention to the specific requirements of the segment.

For Porter, the failure of strategic implementation stems from the lack of strategic discipline, or a firm's inability in pursuing a generic strategy. Since the corporate cultures underlying each generic strategy differs, it is difficult to pursue two strategies effectively. For Porter, therefore, a firm that is "stuck in the middle" is particularly vulnerable to the five competitive forces.

In addition, one has to be attentive to evolutionary processes. Among them are: long run changes in growth, structural change in adjacent industries, entries, and exits. Changes in the long run industry growth is a key factor because it determines the intensity of rivalry and the amount and the pace of expansion necessary to maintain market share. A structural change in an adjacent industry (e.g. the development of a new chain, of network-marketing) may affect industry evolution. Entry and exit affect industry evolution either by adding a new competitor who may well have the skills and resources, or by removing a competitor and leaving a vacuum to be filled.

Current Criticisms: Inherent Fallacies of Formal Planning

Despite all good intentions, the advocates of long term planning were beset by criticisms that it had become a perfunctory exercise that was devoid of meaningful action. Peter Drucker, a renowned management theorist and philosopher, noted that: "In the 25 years after World War II, planning became fashionable. But planning, as commonly practiced, assumes a high degree of continuity. Planning starts out, as a whole, with the trends of yesterdays and projects them into the future — using a different mix perhaps, but with very much the same elements and the same configuration. This is no longer going to work."

In his book, *The Rise and Fall of Strategic Planning,"* Mintzberg has mounted a fierce attack on rational planning and its proponents (Ansoff). He eschews rational planning as unrealistic, if not idealistic, and provides four key fallacies that underlie it.[18] Mintzberg contends that strategy hardly results from deliberation alone. Strategy emerges from a pattern of action and thought. He also sees the strategic process as being overdone, resulting in "process strangulation." Attention is focused on the numeric underlying the process, as opposed to the normative forces that underpin it.

Strategic plans are reduced to perfunctory exercises that force people to make extrapolations of historical numbers. They oblige managers to look back at five years of history and make numerical projections for the next five years and so forth by adjusting for costs, inflation, share, and others. In many organizations, this annual exercise becomes a "fire drill" that everyone does by rote and that no one gives much thought to. These are placed in large folders, often left to gather dust in some bookshelf.[19]

Mintzberg chastises planning advocates as failing to recognize the limitations of their methods. First, planning, by definition, presumes that the external environment can be accurately anticipated and effectively controlled. Yet, such predetermination works fine when the world of planning is stable, thus easing extrapolation and forecasting methods. Contingency methods, such as the much fabled-scenario planning, may also work under uncertain conditions, for which outcomes can still be predicted in terms of probabilities. Even so, in today's current business environments, such conditions do not arise with frequency and predetermination is greatly exaggerated. Secondly, the planning literature preaches "detachment," or the separation of strategic planning from operations management. But the very nature of the hard data that comprise managerial information creates difficulties; much detail is, in fact, lost in the process of aggregation. Consequently, the data used by so-called strategists is faulty and incomplete. Thirdly, it is also a fallacy that strategic planning can be formalized, that systems can detect discontinuities, comprehend stakeholders, provide creativity, and program intuition. Finally, there is the "grand fallacy" that strategic planning is not strategic formulation, because analysis is not synthesis. Analysis cannot ever substitute for synthesis. This is because no amount of elaboration can enable formal procedures to forecast discontinuities, to inform managers who are detached from operations, and to create novel strategies.

Recasting the Controversy: Determinateness Versus Emergent Processes

The debate between Porter and Mintzberg, perhaps two of the most cited authorities in the field, illuminates their differences regarding the strategy process as being deliberate (Porter), or emergent (Mintzberg). Both authors acknowledge that most organizations have developed fairly elaborate strategic planning systems. For Porter, planning systems implant the necessary discipline into the process. When properly employed, strategic planning systems enable managers to understand the dynamics of competition, leading to consistency and focus in decision-making.

Mintzberg argues strategies as plans rarely result as originally conceived or intended. Plans may go unrealized, while patterns may appear without preconceptions. By labeling the first definition *intended* strategy and the second *realized* strategy, then one can distinguish between *deliberate* strategies, where intentions that existed previously were realized, from *emergent* strategies, where patterns developed in the absence of intentions, or despite them. He envisions the strategy process to resemble the art of crafting.

Moreover, Porter and Mintzberg differ on how much strategists can truly discern and anticipate changes in their external environments. The incrementalist school of strategy, represented by Quinn and Wrapp, regards formal planning as important, but as one of many analytical and political events that combine to determine overall strategy. Long range planning is futile (too uncertain), and political behaviors compound rational planning. The actual process used to arrive at a total strategy is usually fragmented, evolutionary, and largely intuitive.

Top managers rarely design their overall strategy in a formal planning cycle, but instead use a series of incremental processes. Pieces of formal strategic analysis do contribute to the final strategy. However, the eventual strategy tends to evolve as internal decisions and external events come together in the minds of managers.

Decisions made incrementally because they tend to be fairly complex and highly uncertain. As a consequence, it is not possible to have the necessary information to engage in formal planning. Managers need time to explore the complexities facing their organization, to experiment with options, and to assess the political consequences of various actions. Finally, managers have to create a widely-shared consensus for action. Quinn's argument is not an abrogation of good management practice. He argues that the rationale behind this kind of process is so powerful that it provides the normative model for strategic decision making, rather than the step-by-step "formal planning systems" approach that is so widely popular.[20]

Quinn's theories appear compelling in situations characterized by high uncertainty and high political activity. Early commitments tend to be relatively open and subject to review, pending more information. Change programs develop in phases with concrete decisions to proceed further only after considerable review by various parties. Final commitments are postponed until the last moment. This is because managers have to sense the need for change, what their employees are capable of, and they also have to be sensitive to politics of the organization. Accordingly, managers need to create the awareness of the need to change, and the involvement of key people. The success of incrementalism depends on top management's ability to create awareness, understanding, acceptance, and commitment need to implement strategies effectively.

For Quinn, strategic decisions are constrained by political boundaries. For managers, it is a timely reminder that it is not only the context of a strategic decision that needs to be managed, but also the processes that ultimately determine the fate of such a decision.

Even so, there is some question regarding the limits of logical incrementalism. At hand is the best way in which to envisage evolutionary change. Incrementalist positions maintain that evolution unfolds in a piecemeal fashion through the accumulation of small changes over time. The more recent view, quantum change, posits that evolution lurches forward between brief moments of metamorphic change that are interspersed between lengthy periods of stability.

If precipitating events jolt the organization, the organizations react incrementally as a way to manage the surrounding aura of uncertainty. Yet, since such jolts, e.g. the oil shocks of 1973 and 1979, affect organizations in different ways, then the pace in which these organizations react — whether it be incremental or quantum — would be an important factor in their future competitiveness.

The discourses on "structure-follows-strategy" amplify the intransigence of structure and processes. The dilemma facing a practising manager becomes clear: to a large extent, events in the external environment indicate what the firm should do, but its own structure and processes limit what it can and cannot do. It becomes essential to understand the dynamics of change, as well as inertial forces within the organization that escalate behavioral persistence over time. In terms of designing organizational templates, the question is whether there are more flexible structures and processes that meet the requirements of changing strategies.

Despite their disagreements, Porter and Mintzberg share the belief that strategies are formulated or crafted by "top" managers.[21] While Mintzberg and his arch-colleague, James Brian Quinn, eschew the distinction between formulation and implementation, they envision strategists as "managing" the process of sensing informational needs, garnering support, and securing commitment.[22] For example, it is Learson who is seen as working behind the scenes in managing the process that led to the adoption of the IBM System 360 architecture. In addition, Porter and Mintzberg believe that good strategy — however formulated or crafted — leads to profit maximization. Porter directly states that the goal of strategy is to establish a sustainable competitive advantage over time — one validated by long term profitability. While not as direct as Porter, Quinn used Fortune 100 companies as exemplars of good strategy decision-making in numerous case studies.

An Unanswered Question: What About Other Cultural Contexts?

For all that is debated, these opposing camps do not critically question the social context that gives meaning to the term "strategy." Both camps view the context underlying their theories as universalistic, or applicable to all competitive settings. Given such, it is not all surprising that strategy is much less extolled in Eastern societies, such as Japan, where the term is ambivalent, or in China, where the term "joss" means a belief in fate, and, by extension, a disregard for outcomes that result from deliberate action.[23]

Much less emphasized in Western thought is the role of institutions in facilitating strategy. Consequently, proponents of such views do not necessarily reside in business schools, but in fragments of political science (John Zysman, Chalmers Johnson) and economics (Kenneth North).[24] Perhaps the closest representative from business side is Kenichi Ohmae who has written a number of books on Japanese strategies.[25]

Box 5-5

The Japanese don't really have a "strategy".
Michael Porter

For institutionalists, strategy matters, but not in the sense that the rational/classicist think.[26] Strategies depend on the particular social settings in which they are embedded, i.e. the extent to which actions can be interpreted and accounted. Strategists deviate from profit maximization deliberately to pursue other outcomes. Strategists deviate from rational thinking, not because they are stupid, but because, in their cultures, the rules make little sense. Firms are not all profit maximizers (as depicted in rational/classic and evolutionary perspectives). Nor are they particularly incrementalists, whose idiosyncrasies are the product of internal limits and political compromise. Norms derive not so much from cognitive bounds but from the cultural rules of society.

Box 5-6

> **Western managers...need to get out of the old mode of thinking that knowledge can be acquired, taught, and trained through manuals, books, or lectures. Instead, they need to pay more attention to the less formal and systemic side of knowledge and start focusing on highly subjective insights, intuitions, and hunches that are gained through the use of metaphors, pictures or experiences.**
>
> Ikujiro Nonaka and Hirotaka Takeuchi, *The Knowledge Creating Company*

It is not at all surprising to institutionalists that the Japanese, for example, devote little attention to strategy. After all, strategy, as conceived by the rational school, does not make much sense in this culture. It is more likely for the Japanese to attribute strategy as a product of relational activities between politicians, government bureaucracies (such as the Ministry of Trade and Industry), and business firms. The *keiretsu*-like groupings, so unlike those in the Western world, provide incentives (i.e. bank financing, interfirm sharing of directors and funds, contractor-subcontractor relations) that compel Japanese to act in certain ways that Westerners find difficult to understand (why invest resources during recessionary periods), or controversial (why sacrifice profitability for market share).[27] Embedded in these explanations is a theoretical bias that favors the experiences of Western firms. In fundamental terms, this bias is manifested in explanations for alleged Japanese opportunistic firm behaviors that are perceived to be contrary, if not antithetical to conventional "Western" theories and perspectives.

Imperatives for a New Mind Set

The First Industrial Revolution is traced to the end of the eighteenth century with its origin in Great Britain. The trigger for this revolution was the invention of the spinning wheel and mass communication. The emergence of new technology at this time enabled capitalists to locate labor into central factories. Flushed with a high supply of labor, this was not a problem. Eventually, this ushered in the growth of small to medium sized enterprises, with the state playing a non-interference role, i.e. *laissez faire*.

The Second Industrial Revolution, which occurred in Germany and the United States about 100 years later, shared the distinction of generating new products and processes. New technology enabled Germany and the United States to overtake the economy of Great Britain. Since more sizable investments in research and development were required, the center of development became the multidivisional corporation. And, because of the migration of skilled workers from Europe, labor again was in abundant supply. The State played a greater role in enacting infant-protection measures to aid fledgling corporations and industries.

Late industrialization is the central topic of study by Alice Amsden.[28] If industrialization first occurred in England on the basis of innovation, and if occurred in Germany and the United States on the basis of innovation, then it occurred among the "backward" countries on the basis of learning. Innovators are aided in the conquest of markets by novel products and services. Learners, by definition, do not innovate and must compete initially on the basis of low wages, state subsidies, and incremental productivity and quality improvements related to existing products. In turn, different modes of competing are associated with differences in the firms' strategic focus.

In the U.S. economy over the 100-year span from 1880 to 1980, three major periods are easily identified, and the transition to a fourth is currently underway. Each of these periods begins with organizations facing large transitions in their external environments — a combination of social, political, technological, and market challenges. Within each period, organizations have developed new strategies to respond to the new environment and to develop organizational structures and processes to make strategies work. Historically, only a few companies led the way. Through early tight fit, these firms enjoyed enormous success and became hallmarks of the period.

Technology is the force that combines the tenets of late industrialization and restructuring of conglomerates. Already, we are witness to the new requirements of this emerging organization. As defined by William Davidow and Michael Malone in their book, *The Virtual Corporation*, this new organization should have the capability to offer products and services that are produced instantaneously and customized in response to customer demand.[29] To achieve this, companies have proceeded to flatten their structures, reengineer their processes, and develop highly flexible responses to customer needs through information technology and network organizations.

In the next decade, strategic implementation will become increasingly complex, even surpassing the capabilities of seasoned managers to develop quick and timely decisions. In the early part of the century, implementation, then conceived as separate from strategic formulation, was defined as a "fit" between strategy and structure. Strategies dictated the appropriateness of organizational structures. Michael Porter attempted to refine this concept in suggesting strategic "discipline." Implementation would be based on a careful appraisal of a firm's competencies through the value chain. Successful firms were those that avoided getting "stuck-in-the-middle"; effective strategies flowed from disciplined implementation. Richard D'Aveni argues that this is no longer sustainable over time, and that firms had to learn how to disrupt the status quo and to develop initiatives aimed at creating temporary advantage. Raymond Miles and Charles Snow provide the imperatives for three types of network organizations that facilitate this constant jockeying for strategic position.[30]

Evidently, the new mind set for operating in this environment draws its foundations from different foundations that characterized the first two industrial revolutions. We argue that these new foundations have already taken place, and these are grounded in the movement from Newtonian to Quantum Thinking, from Cartesian Division to Holistic Patterns, from Equilibrium to ordered Chaos, and New Growth Model

Theory. These have been discussed in our first chapter. As a point of transition, we now relate these to new strategic thinking.

New Concepts: Core Competence and Strategic Intent

Newer conceptions of strategic planning have focused on the argument that corporate as well as competitive advantage is based on the unique resources of the firm, and the way these resources are deployed.[31] Resources are critical because they define what the firm can do, and distinguish one firm from another. The most valuable resources are those that enable a firm to compete successfully in many markets, such as Microsoft's capabilities in operating systems that have broad applicability across many industry sectors.

In 1990, C.K. Prahalad and Gary Hamel introduced the notion of "core competence" that has become highly influential among strategy thinkers.[32] A core competency, as described by them, refers to capability or a skill that provided the cohesive thread through a firm's business. To the extent that these core resources are scarce, unique, and difficult to imitate, then these can sustain the firm's competitive position and can leverage this position into other lucrative businesses.[33] Collins and Montgomery have enlarged this notion in terms of "organizational capabilities" to describe complex combinations of assets, people, and processes used by organizations to transform inputs into outputs. Other researchers have honed into this concept, distinguishing between static routines (competencies) that embody the capability of replicating previously performed tasks, to dynamic routines (competencies) that enable a firm to develop and adapt further to changing competitive demands.[34]

Box 5-7

If you want to escape the gravitational pull of the past, you have to be willing to challenge your own orthodoxies. To regenerate your core strategies and rethink your most fundamental assumptions about how you are going to compete.

C.K. Prahalad

Reflecting further on the notion of core competencies, C.K. Prahalad and Gary Hamel argue forcefully for a configuration of an organization as a portfolio of core competencies, rather than simply a portfolio of business units. They argue that business units are focused on products and markets, while core competencies are focused on consumer benefits.[35] Apple's "user-friendliness" and Sony's "pocketability" are offered as examples. Through the proper identification and understanding of its core competencies, a firm can become synergistic and properly mobilize the energies of a firm towards a desired future state. This is facilitated when a company develops a strategic intent, that is, a widely shared aspiration to have a goal that is clear, cohesive, and an obsession with winning. Thus, a strategic architecture, as is one embodied with competencies, is not enough; a firm needs a shared aspiration that provides its emotional and intellectual energy to make the journey.

Is *Hypercompetition* the Answer?

While the preceding treatise about core competencies and strategic intent are directed at sustaining competitive advantage, Richard D'Aveni takes on the challenge in explaining why traditional sources of advantage no longer provide long-term security.[36] One does not have to look beyond the experiences of IBM and General Motors, household names in computers and automobiles, that were once viewed as unassailable in global competition. Both companies had economies of scale, massive advertising budgets, excellent distribution systems, cutting-edge R&D, deep cash pockets, and power over buyers and supplies. Yet, both fell prey to competition, both companies appeared inertial, and unable to exploit major opportunities. Hewlett Packard CEO Lewis Platt says that, "The only mistake they (the above companies) made is, they did whatever it was that made them leaders a little too long."

In D'Aveni's' view, hypercompetition results from a series of competitive and countermoves, that is, 'dynamic strategic interactions' that lead to the erosion, destruction and neutralization of each firm's competitive advantages. These dynamic strategic interactions occur in four areas: (1) cost and quality competition, (2) timing and know-how competition, (3) competition for the creation and destruction of strongholds, and (4) competition for the accumulation and neutralization of deep pockets. Competitive moves escalate with such ferocity that traditional sources of competitive advantage can no longer be sustained.

Porter and his colleagues stress the pursuit of a sustainable competitive advantage. Consider strategy-guru, Pankaj Ghemawat, who stated: "The distinction between contestable and sustainable advantage is a matter of degree. Sustainability is greatest when based on several kinds of advantages rather than one, when the advantage is large and when few environmental threats to it existed."[37] D'Aveni argues that strategy is also the creative destruction of the opponents' advantage. He views hypercompetition as particularly pervasive, extending from high technology industries to more mundane ones like hot sauce and cat-food.

D'Aveni's research suggests that the escalation of competition occurs *within* each area (above) or *across* the arenas. For example, firms escalate by increasing the level of quality or lowering the price of their goods. They also escalate efforts to develop new know-how, move faster, invade or create new strongholds and build deep pockets. Competition continues until firms exhaust the advantages of that arena. Then, they move on to know how in the second arena, until the benefits of these advantages become too expensive. A third possible step is to attempt to create strongholds to limit competition, until these too are finally breached. This leads to the use of deep pockets until which time firms deplete their resources or join alliances to balance off the resources of competing alliances. While such a progressive ultimately finds resolution in a 'perfectly competitive' world (at least, theoretically), this state is purposefully avoided by corporations seeking profit. In one sense, a paradox occurs: while firms act in ways oriented towards achieving perfect competition, they must attempt to avoid it in order to attain abnormal profits, thus leaving them to hibernate in a hypercompetitive world.

Drivers of Hypercompetition

Why does hypercompetition exist? What are the 'drivers' of hypercompetition? D'Aveni identifies the following:

Growing Instability of the Environment. In the past, some companies successfully avoided intense competition through implicit collusion or developing sustainable advantages. Then, led to stable 'equilibrium-seeking' conditions (note the stability of oligopolistic competition, for example). Current environment makes collusion a less viable alternative. Moreover, new entrants, particularly foreign competitors, are not compelled to 'play by the rules'.

Erosion of Trust. This is a result of the evolution of global markets, new technologies and maverick modes of competition. Cross-cultural distances compound communications; and, this in turn, lead to lack of trust, even opportunism. This is also described as a "prisoner's dilemma", where both firms would be better off if they cooperated, but neither side can trust the other to sustain the agreement.

Escalating Commitments. The escalation into hypercompetition, in some ways, is similar to arms race between two countries. The negative effects are not initially seen nor anticipated. Firms make short term decisions to make temporary advantages, unaware that competitive countermoves act to 'up the ante'. Once the dynamics are in motion, the players don't know how to stop it (re: The airline price competition in recessionary periods). Competition is heightened, even though it makes more sense to cooperate.

Strategic Implications of Hypercompetition

What are the implications of hypercompetition? New rules of competition dictate fundamental changes in premises. Success in a hypercompetitive market is based on a number of paradoxes:

Destroy competitive advantages to gain advantage. Since every advantage eventually is outmaneuvered, companies are forced to destroy their own competitive advantages to create new ones. The challenge is for companies to get the most out of existing advantages before destroying them to create new ones.

Entry barriers only work if others respect them. Firms can deter competitors from entering their markets only if the competitors do not want to enter the market, in which case the entry barrier wasn't a barrier at all.

A logical approach is to be unpredictable and irrational. Too much consistency and logical thinking can make a firm predictable. So, the company must at least appear to be irrational, but not crazy.

Traditional long-term planning does not prepare for the long term. Long-term success depends not on a static, long-term strategy but on a dynamic strategy that allows for a series of short-term advantages.

Attack a competitor's strength, not the weakness. Using the company's strength against an opponent's weaknesses, may work once or twice, but not over several iterations. Over time, the competitor works hard enough to develop a weakness into a strength.

Compete to win, but in doing so, the stakes are raised. Each move up the escalation ladder raises the stakes of the game and makes winning more difficult.

D'Aveni's studies of successful and unsuccessful companies in hyper-competitive environments reveal seven key elements of a dynamic approach to strategy that encompass the three factors for effective delivery of a series of market disruptions: *vision, capabilities,* and *tactics.* The New 7-S framework is based on a strategy of finding and building temporary advantages through market disruption rather than sustaining advantage and perpetuating an equilibrium. The 7-S model addresses all three factors for effective delivery of a series of disruptions, providing increased emphasis on the first two levels (vision and capabilities) and more creative approaches (tactics) than many firms currently use.

Is *Hypercompetition* the answer? As compelling as D'Aveni's arguments are, two of his premises are not readily resolved: (a) the nature of the environment that is producing hypercompetition, and (b) the ability of firms to learn quickly enough to be able to respond in a manner suggested by their strategic interactions. D'Aveni argues that the competitive environment is primarily created by competitive interactions, more specifically, moves and countermoves by competitive actors. But, why would a firm pursue strategies that would lead to hypercompetition, when a steady-state would benefit not only the firm but all others in the industry? One can assume that firms, aware of the consequences of their actions, will avoid those that are self-destructing. Even maverick firms will avoid self-inflicting wounds, that is, they will assume the risks of escalating competition only if they anticipate that the gains will far exceed their losses.

We argue that behavior in hypercompetitive markets will depend on the learning abilities of organizations, specifically their perceptions of "win-loss" within a changing and unclear environment. It may also depend on external factors, like national priorities or corporate egos. It may even depend on human nature. Throughout history, we note that a high proportion of the nations that started wars wound up losing the wars they started. That is not logical, but it is true.

In earlier chapters, we have argued that the application of quantum physics, holistic patterns, ordered chaos, and new growth theory provide the basis for the pattern of growth and renewal that undergrids this new business environment. Because this type of thinking is relatively new, it is not as well ingrained in conventional theories. Nevertheless, it provides the setting for which D'Aveni's *Hypercompetition* becomes a viable reality. In the next section, we offer a comparison between traditional and contemporary approaches to the study of corporate strategy. These form the basis for the next chapter where we discuss how the Engines of Prosperity, or patterns of growth and renewal, define the growth trajectories of Information Age industries.

Comparing Traditional and Contemporary Approaches to Strategy

Box 5-8

> **You can't look at the future as a continuation of the past...because the future is going to be different. And we really have to unlearn the way we have dealt with the past in order to deal with the future.**
>
> Charles Handy

In Exhibit 5-1, we depict traditional and contemporary approaches to strategic planning, as developed in the preceding sections. Taking classical strategic precepts as our point of departure, we demarcate generic strategic planning in six phases: corporate mission, external and internal analysis, strategic alternatives, sources of competitive advantages, and implementation. Even with this rather simplistic construction, important differences between the schools of thought can be described and delineated.

Corporate Mission

To start with, an important hallmark of strategic planning is the "mission" of a company, defined loosely as the definition of the firm's market or industry, the customers it seeks to serve, and its distinctive approach to its business that takes into account any important values and philosophy. As postulated by classical theorists, a firm's mission is what guides, directs, and even inspires its actions and behavior. While the mission demarcates what the firms strives to do and how to do it, it is not mistaken for specific goals or strategies. It is anticipated that goals and strategies may change depending on a firm's response to its external environment, but that the mission, or the underlying vision that may be embedded in it, remains constant.

Exhibit 5-1: Comparing Corporate Strategy: Traditional (Capital-Intensive) and Contemporary (Knowledge-Based).

GENERIC STRATEGIC PLANNING	*TRADITIONAL APPROACH*	*CONTEMPORARY APPROACH*
1. Corporate Mission	Examination of the firm's primary purpose, i.e. where we are and where we should be.	Appraisal of the firm's field, i.e. employees, management, stakeholders, etc. who affect the firm's activities.
2. External Analysis	Opportunities and threats as identified from industry dynamics and competitive behavior.	Dynamic analysis that examines changes in industry boundaries.
3. Internal Analysis	Examination of corporate values, competencies, and philosophy.	Emphasis on a firm's strategic architecture, core competencies, and strategic intent.
4. Strategic Alternatives	Systematic procedures for selecting among various strategic alternatives relative to competitive position to meet requirements of the mission.	Systematic procedures directed at "Which alternative will force the structural changes and result in a state of a system defined by the mission statement?"
5. Implementation	Congruence, or co-alignment, between environment, structure, and processes.	Networks, or similar forms of *ad hoc* virtual structures, that are consistent with the unbundling of a firm's value chain.

Box 5-8

The future is not what it used to be.

Arthur C. Clarke

In classical and traditional approaches to strategy, a company's mission is often developed and formulated in linear, mechanistic terms.[38] In other words, managers are trained to think in terms of destinations as future desired states. Managers are taught that the more clear the vision of this future, the more force it exerts on the present. In many ways, it has a strong Newtonian underpinning, much like the old view of gravity.

In this Information Age thinking, the concept of mission changes from unidirectional polarities (i.e. "to become the market leader in air purifiers") to that of a field. The formative properties of this field consists of all employees and stakeholders who are likely to define and change this field. Thus, the future of an air purifier company is likely to be determined by its management, its employees, government agencies, and leading-edge technologies. Instead of thinking of a mission in linear terms, a company's desired future then becomes a product of how inhabitants of the field work together, develop consensual goals, and resolve conflict between them. Energy, synergy, or inertia become important factors defining this company's mission. Instead of a linear trajectory to define mission, its focus is more horizontal, with attention placed on the actions and behavior of cooperating or competing stakeholders.

External/Internal Analysis

The assessment of a firm's mission is preceded by an appraisal of external and internal environments; opportunities and threats are typically identified after a thorough industry analysis, while strengths and weaknesses are defined from a firm's competencies. Strengths and weaknesses are relative to the firm's previous abilities, or to the abilities of the firm's competitors. In classical theory, it was essential to recognize factors with marginal improvements in order to encourage the development of these abilities, and to intervene in the case of declining ability in order to arrest what is then anticipated to be the start of deteriorating performance.[39]

Contemporary approaches to external analysis focus on the permeability of industry boundaries, oftentimes eschewing the very notion of an "industry classification." C.K. Prahalad notes that the traditional view of competition is based on assumptions that industry boundaries are clear and that industries have distinctive characteristics.

He argues that "there is absolutely no way, in the evolving marketplace, that you can know exactly who are the suppliers, customers, competitors, and collaborators."[40] Moreover, as industries merge and commingle, it is no longer clear what the product is or what its value is to the customer. The fusion of computers, consumer electronics, communications, entertainment, etc. provides one example where industry boundaries are no longer meaningful.

Gary Hamel shares this same view, and he even states that he no longer uses the word "industry" because it is not useful. In more and more industries, such as multimedia or financial services, the "five force analysis" cannot be done at a broad level of analysis, and even when done at a narrow level, there is the risk that new opportunities might not be identified. As examples, Hamel uses Coca Cola. If the cost of the product is X and the price Y, it is difficult to assess who gets the difference. It can be argued that the manufacturer of the aluminum can, the supermarket that retails the product, or the company that transported the goods all can rightfully claim it.[41]

In terms of internal analysis, which was once dominated by simple notions such as gap analysis, more contemporary approaches have focused on strategic architectures, core competencies, and strategic intent. Since this is developed at length in the chapter, only the main conclusions are reiterated in this section: core competencies and dynamic capabilities are more meaningful in capturing the essence of rapidly-changing industries than strategic business units that are more oriented to static, capital-intensive industries.

Strategic Alternatives

Following the assessment phases, the classical strategic model would then take a rational sequence in defining the strategic issue, the nature of choices, preference functions for deciding on outcomes, and implementation guidelines. These assessments are designed to provide closure to how a firm can best realize its mission. They would also address the structural changes needed to attain the desired future state.[42]

Recall, however, that contemporary representations of mission statements are more oriented at horizontal, multidimensional assessments, rather than linear extrapolations. Because outcomes tend to be multidimensional, they define and restrict the future course of the firm to specific trajectories that lead to a desired state. Therefore, the question should not be: "Which alternatives might be consistent with some general mission statement and result in the attainment of some lesser goal?' but "Which alternatives will force the structural changes and result in a state of a system defined by the mission statement?"[43]

Implementation

Implementation can take on a different character when firms compete on the basis of knowledge and technology as opposed to factor inputs and production. Classical strategic thinking, and as such has been extended to mainstream management thought, focuses on congruence as a basis for organizational balance, order, and harmony. The concept is powerful as it is in accord with accepted economic notions of equilibrium. As argued earlier, even seemingly disorderly processes, such as logical incrementalism, has sequential fit as its basis for reducing uncertainty and in garnering political consensus.

The emerging organizational structure for Information Age firms would appear to conform more to constellations of organizational networks. Instead of "congruence" as posited in classical/mainstream theories, the driving principle will be that of "creative outsourcing." The antecedents of unbundling relate back to the so-called "virtual organization," defined as a firm with the capability to offer products and services that are produced instantaneously and customized in response to customer demand.[44] For example, in a treatise about knowledge-based services and technology-based firms, Quinn describes how selected companies develop strategies around core competencies, while outsourcing less important functions to superior outside vendors.[45] As firms select which competencies to retain, and which activities to outsource, various configurations of structure emerge.

Relatedly, Handy's conception of a "shamrock" organization consists of three-layers of employees: the core workers (professional core), contractual workers (work that is outsourced), and the flexible labor force (those who are part-time or temporary) runs external to the organization. How work is divided depends on what one is willing to pay for use of a service and on how much the service provider or unit is willing to accept to render the service.[46]

In most knowledge-based organizations, the cost of services is allocated to the revenue producing units in the external service delivery. These allocations are frequently based on budgeted or actual costs, and such costs act as a representation of the stability of value relationships among employees. Such relationships gain stability as behaviors or activities are regulated by cognitive (not just financial) assessment of value or importance. If the work was critical or of high value, you might pay a premium to get the best, most trusted talent available. If it was mundane, you might use minimum wage temporary workers.

For example, given a choice, most people would not pick a brain surgeon based only on who gave the cheapest bid. In a similar manner, Information Age firms may pick more expensive consultants or suppliers who bundle special services, value, or knowledge as a part of their product offering. Conversely, Machine Age firms, particularly those stressed by downsizing, tend to view supplied components and employees as generic and interchangeable. These firms don't typically apply a lot of cognitive energy to sourcing selection. Their strategy is usually set by process to something like "minimum acceptable" and "cheapest."

Recommended Action

This chapter has given a brief overview of current thought on strategy in a business context. In the end, strategy discussions take on an almost religious overtones. Some firms and cultures prefer one type of strategy, some choose others, and a few actively study them all. In fact, when an industry slides into a condition where the major players all use the same strategy, it often poses opportunity for "outsiders" with different belief sets.

Clearly, no one strategy works well in all markets or against all opponents. Still, one hopes that the major strategies have all proven useful sometimes, against some opponents. In the end, knowing when to apply which strategy is a matter of skill, judgment, and experience.

As professionals, the authors study all the strategies. We suggest that such strategic understanding should be viewed as "core knowledge" for managers. All the major strategies should be understood, and those preferred by one's likely competitors should be understood in detail.

Microsoft, to use them again as an example, carries this down to the individual opponent level. If engaged in major battles, Microsoft's market managers keep pictures of their individual opponents and their families posted on their walls. They make the effort to understand how these people are inclined to think. If they are losing ground, they may add pictures and details. If winning, they may "celebrate" by removing a photo for each victory.

One of the authors asked one of Japan's MITI officials, why they were so intense about sending their managers to U.S. business schools. If they were leading in so many markets, what could they hope to learn that would be worth the investment? The official replied, "We can learn how you are trained to think, and this is useful knowledge. It helps us predict how you will act." Military commanders, like World War II General Patton who studied his opponent Rommel in detail, understand the MITI or Microsoft approach fully.

Chapter 6

The Engines of Prosperity: Applications

Key Themes at a Glance

In our view, the later stages of the Third Industrial Revolution are characterized by industrial transformations brought about by the capability of digital technology. Advances in technology have created new patterns of growth and renewal.

Previously stable environments are becoming hyper-competitive, often as a result of abrupt shocks as they undergo periods of rapid transformation and intense competition. Contrary to the predictions of linear life cycle models, some mature industries are undergoing a "de-maturity" process, and once again have become hotbeds of innovation and change.

The roots of this industrial upheaval are grounded in five basic factors, which we call the Engines of Prosperity: *steep learning/cost curves, demand amplification, bandwagons and "lock-ins", innovations and disruptions,* and *the unbundling of value chains.* As applied to certain knowledge-based industries, the resulting pattern is one of increasing returns to scale, and not the decreasing returns of scale as is customarily conceived of many other industries.

Two case examples illustrate the engines in context: videocassette recorders (VCRs) and semiconductors. As technological advances permeate VCRs and semiconductors, new applications are developed, traditional industrial boundaries are blurred, new designs replace traditional ones, and new networks of industry communities evolve.

And as competition intensifies, so does the uncertainty that pervades competing firms. It is essential that firms understand the dynamics of growth, development, and renewal as exemplified in the Engines of Prosperity. The new rules transcend agility, speed, and discipline. They call for establishing value in the context of a seemingly endless cycle of cost reduction and increased market demand.

America is in the throes of a new revolution. It is not a political upheaval of democratic traditions. Nor is it an engagement of contestable ideologies. It is something far more latent, but with far reaching implications: *it is a revolution of industrial transformations brought about by the capability of digital technology.*

Consider the pricing decision once faced by Computer Associates International Inc upon the introduction of an accounting program, *Simple Money*, in 1993.[1] It had priced its first million copies at *zero*. Cellular phones can be priced as low as $0.01. Toshiba Corporation happily admits that its forthcoming digital-movie player will probably never earn back its initial investment. Teleport Communications Group, Inc and Teleport crew will install a dozen optical fibers with a million times the capacity for no extra charge. What is it about the current information technology environment that leads firms like Computer Associates and Toshiba to "give products away"?

In these cases, the logic is that businesses can thrive at the very moment their prices are falling the fastest. Computer Associates would give away its software on the premise that word-of-mouth might far outweigh the trivial costs of making software — and persuade customers to buy upgraded programs. Cellular phones are given away or sold for almost nothing to secure customers for more expensive monthly services.

The concept of selling below "cost" is hardly new. The difference is that the new Information Age economic dynamics — the Engines of Prosperity — make it much more appealing and widespread. Everyone knew that "You give away razors to sell razor blades," but a product manager at IBM in the 80s who recommended giving products away free would probably not keep his job. Indeed, it was not uncommon in that era for prices to be set, and raised each quarter, by corporate edict.

Times have changed. Today, Microsoft cheerfully gives its web browsers away free to attack Netscape's dominant market share. Netscape, who got that share with the same strategy, is relatively less affluent and now must charge for their browsers. Toshiba intends to recoup its developmental costs for digital videodisks with spin-offs to other areas such as high capacity audio players to storage devices for laptop computers. Optical fibers are so cheap that it makes sense to install them for a lifetime.

The rapid pace of globalization has induced business scholars and practitioners to reassess their previously held beliefs about competitive rivalry and the need for cooperation. The traditional basis of competitive rivalry can be neatly classified in terms of generic strategies such as cost leadership, differentiation, and focus. Moreover, one was able to associate particular types of strategy with specific types of market conditions.

For example, differentiation was associated with emerging products for which market uses had still to be defined. Mature products, defined by their commodity-status, were subjected to cost leadership type of competition.[2]

The distinction between the two in the context of high technology products has decreased and is no longer very compelling. Most electronic products have become "high tech commodities" in that they combine characteristics of mass production with extremely short product life cycles and periodic trajectory-disturbing innovations. Semiconductor manufacturers, for example, have had to contend with simultaneous needs to develop an innovative new stream of products, while, at the same time, manage a standardized product well into maturity and end of life stages. As a result, cost competition has to be combined with differentiation and speed-to-market.[3]

This is either a vicious or virtuous cycle, depending on whether you are winning or losing. New products flow into markets as a series of rhythmic waves. The new, innovative products capture the share of market from the old products they replace. Prices for the old product collapse as they become commodities. Global competition gives a new meaning to the word commodity, as the old products can be built with slave labor and near zero margins almost indefinitely, a fact which makes exit strategies as challenging as entry strategies. Soon the new products become old and are replaced, or are cloned, and the cycle repeats.

Periodic trajectory-disrupting technologies have also dismantled entry barriers and have eroded competitive advantages.[4] At the same time, rapid technical advances have produced very steep entry barriers (that is, microprocessors). Governments affect global competition through a variety of regulatory barriers that restrict access to new technologies, product standards, and markets. Thus, entry deterrence strategies become crucial. Microsoft's great wealth has come largely from the entry barriers it has built into PC operating systems and the core applications which run on them.

To compete effectively on a global scale, a firm must be able to internalize on a global scale, specialized assets, technological knowledge, organizational competence, finance, production, supplier and customer networks, and market intelligence. Of equal importance, a firm has to build these competencies much quicker and at less cost than its competitors.[5]

The confluence of change that moves us from the Machine Age to the Information Age is the basis of this new competition. The earth has become a global village, allowing for turbulent and fast changing worldwide markets. New knowledge — the learning curve and the demand curve — makes time crucially important.

The iron laws of steep demand and learning curves foredoom those unfortunates who lag. Unless your competition makes major mistakes — and few of the leaders make such blunders — low volume positions in products or markets are very difficult to defend. That creates constant pressure to invent new products and markets to negate the advantages of your competitors. If you don't respond to these pressures, you find that your pricing flexibility, profits, margins, and market share can evaporate very quickly.

More significant, the dynamic power of the Engines of Prosperity is so steep and the laws which govern them are so powerful, that once you get "behind on the curve" it is all but impossible to recover. Mighty IBM, with a technically superior operating system (OS/2), got behind Microsoft and never recovered. It cost them over a billion dollars and relegated them to a minor position in the PC industry.

Knowledge comes from information and experience. The contenders in the dynamo markets keep teams intact and cycle products rapidly so that they can learn faster than competitors and enjoy the rewards of market share and production volume. Winners develop endless streams of ever better products for their existing markets, to hold and improve their positions. Winners also continually probe new markets to develop "headroom" that is — at least for a time — free from competitors. Any firm from any nation can win with the right products, but the dynamo markets cycle relentlessly. Victory in one cycle only wins the opportunity to compete in the next, but missing a cycle can be death.

In this chapter, we specifically discuss the drivers of this new competitive environment. These drivers explain why Computer Associates, Toshiba, and Cellular are willing to give away products because they anticipate much larger returns down the road. Taken altogether, new rules for operating in this environment transcend agility, speed, and discipline. They call for redefining value in an economy where the cost of raw technology is plummeting to zero, and where the value is in establishing a good will relationship with present customers. The seemingly endless cycle of cost reduction and increased market demand creates an era where the only thing that matters is that the exponential growth of the market is faster than the exponential decline of prices. This inverted logic results from rapid technological advances that lead to lower prices, which, in turn, spur market demand.

We describe this new environment in terms of five basic components that are termed *Engines of Prosperity* to denote their generative power in developing high growth and renewal. The first three engines relate to environmental conditions; the next two relate to distinctive responses by firms to cope with these environmental contingencies.

Engine #1: Steep Learning/Cost Curves

Although the scientific and industrial origins of information technologies can be traced to earlier decades (re: the invention of the telephone by Bell in 1876, the radio by Marconi in 1898, and the vacuum tube by De Forest in 1906), it was during World War II and its aftermath that major technological breakthroughs in electronics took place: the first programmable computer, the transistor in 1947, and the integrated circuit in 1957 by Jack Kilby and Robert Noyce. The significance of these innovations is felt most markedly in how costs and prices have spiraled downward.

It only took three years for the price of semiconductors to drop by 85 percent and in the next ten years, production increased twenty times.[6] As a point of comparison, it took 70 years (1780–1850) for the price of cotton to drop by 8 percent during the Industrial Revolution. During the 1960s, better chip design and more efficient manufacturing methods cut the average price of the integrated circuit from $50 in 1962 to $1 in 1971. The giant leap forward came in 1971 with the invention of the microprocessor by Ted Huff. On one estimate, computer power now costs only one-hundredth of 1% of what it did in the early 1970s.[7] Never before in the history of industrial applications have we seen such a dramatic fall in prices, and this trend is continuing.

Most industries are subject to a fairly common and well-understood premise: as one produces more and more of the same product, the costs of production fall at some proportional rate. Therefore, the cost of producing the second automobile, for example, is less than producing the first, and so on. The formal application of this principle is traced back to the manufacture of airplanes during the late 1930s and Liberty ships during the Second World War. There was a remarkable regularity in the reduction of costs and prices when production volume doubled.

The experience curve was discovered in 1925, and the Commander of the Wright Patterson Air Force Base is given credit. It says that the more of something you make, the cheaper it gets. The curve is logarithmic, so to observe the effect you need to observe a huge volume of purchases.[8] Best case learning curves for Machine Age products are in the 80 percent range. That means that each time you double the number of samples, cost is reduced by 20 percent. If the first airplane cost $10 million, the second would cost $8 million, the fifth $6.4 million, etc. Even during World War II, the military used this rule to set cost targets.

Several reasons account for cost reduction: defects are eliminated with each iteration; more efficient processes can be identified and introduced at later stages; workers become increasingly familiar with the production processes. In these cases, learning takes place. The important thing is that the improvement continues across the entire curve. For an 80 percent learning curve, you get the same 20% cost reduction between the one millionth and two millionth unit as you did between the first and the second.

Industries like agriculture have very flat learning curves. The cost decline between the one millionth and two millionth apple may be barely detectable. Information Age industries, conversely, have phenomenally steep learning curves. The development cost of making the first copy of a new microprocessor may be over a billion dollars, while making the second copy could be almost free. That is, in part, why we can expect increased intellectual property theft and major international "patent wars." Those who invest millions of dollars, or more, to create innovative products and businesses need strong legal protection for their knowledge if they are to benefit commercially.

In his 1890 *Principles of Economics*, the Victorian economist Alfred Marshall noted that if a firm's production costs fell as their market shares increased, a firm that simply by good fortune gained a high proportion of the market early would be able to sustain an advantage over its rivals. Yet, it was not until the 1970s that the world of business discovered how to exploit the advantages of such economies as an integral part of corporate strategy. One consulting firm, Boston Consulting Group (BCG) used — some say misused — it to build a prosperous worldwide practice. They taught clients to compete based on market share. BCG's concept was simple. If you had more market share, then you built and sold more units than your competition. Since each unit and each activity cost you less than it did your competitors, you could enjoy superior profit margins or sell for less and gain share.

While results were mixed, BCG's logic was sound. Not surprisingly, the Japanese and Pacific Rim competitors adapted better to strategies of sacrificing profits to gain share than did the West. Today most agree that trying to defend a low share position against competent competitors is very difficult. The hyperbole in academic circles is that new technologies are subject to steep learning curves, that is, disproportionately large cost reductions corresponding to increases in output. This applies to certain products, but not others.

Part of the economy that is resource-based (agriculture, bulk-goods, mining) is still for the most part subject to diminishing returns.[9] But Information Age industries like software and semiconductors can have very steep learning curves. Falling prices is one clear indicator of technological progress. Combined with rising affluence of customer groups, lower prices accelerate the speed of adoption and diffusion. This sharp decrease in costs and prices is illustrated in Exhibit 6-1. Even so, unlike the scenario suggested in conventional treatises, learning curves do not necessarily result in maturing products that are then eased into oblivion in favor of newer products. The second piece of the argument is what happens to consumer demand as innovations become more diffused and as costs and prices continue to fall.

Engine #2: Demand Amplification/Increasing Returns

The second engine relates to a shift in our assumptions about cost patterns over time. Conventional theories of competition are built on the assumption of diminishing returns. The assumption is that product

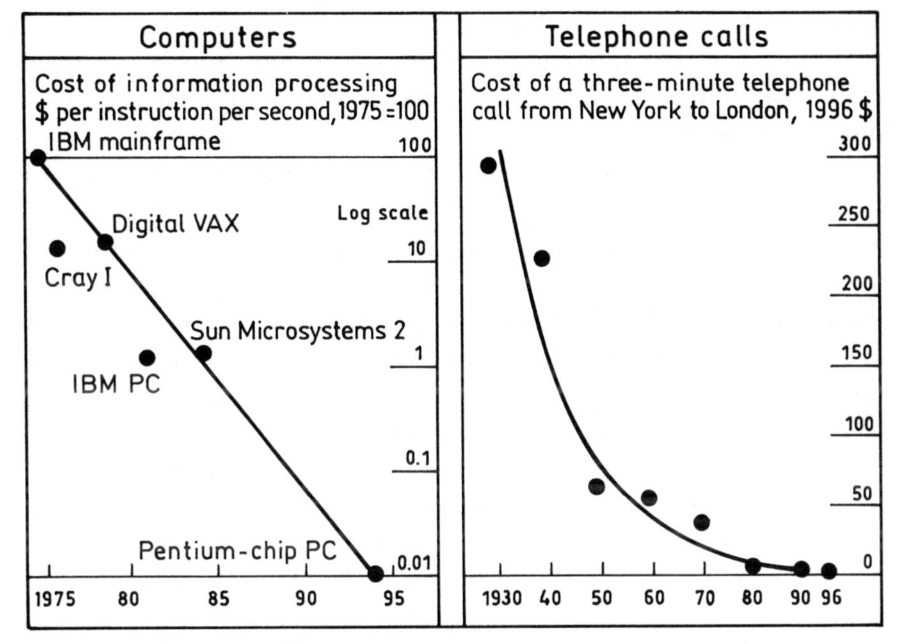

*Data from World Bank, IMF
SOURCE: "Taking the Plunge," The Economist, October 18, 1997.

Exhibit 6-1: "Taking the Plunge": Illustrating Steep Learning Curves.

usage, however attractive, is limited. As usage of a product increases, the marginal benefits associated with it eventually start to decrease. Economic actions trigger feedback that leads to predictable equilibrium for prices and market shares. Feedback stabilizes the economy because the major change will be offset by the very reactions that they generate. Equilibrium marks the "best" outcome possible under the circumstances: the efficient use of resources. Diminishing returns to scale implies a single equilibrium point.[10]

A different conception applies in the case of increasing returns to scale. Positive feedback magnifies the effects of small economic shifts, resulting in multiple equilibrium points. Increasing returns implies that benefits amplify with increased product usage. Increasing returns to scale entails multiple equilibrium points since there is no guarantee that the particular outcome selected among the many alternatives will be the "best" one. Predictably, shared markets are no longer guaranteed.

Information Age markets tend to be "winner takes all." This is because of increasing returns to scale. For example, Microsoft's Windows has, as a practical matter, driven out all alternatives for desktop operating systems. To an even greater extent, a case we'll discuss in detail, so has the VHS videotape format. Conversely, Machine Age markets rarely tended to be that way, unless someone had monopoly control of a crucial resource. No matter how successful, say, a Volkswagen or a Corvette or a Taurus was in the market, it was hardly conceivable that it would ever drive out all other cars.

The concept of demand curves had been around in consumer circles for decades. It was less of a science, but everyone knew that if you dropped prices more people would purchase. Most products' demand curves are flat. Let's say, for that sake of argument, that these curves are typically in the 80% range. If you cut the price of a product in half, then 20 percent more would be sold. That is not very appealing, since it probably yields negative profit margins. Indeed, whole industries were turned unprofitable by BCG's market share driven strategy run amok.

Ironically, as early as the 1930s, Oxford's economist Allyn Young, was speculating that there might be product-market areas that could lead to limitlessly increasing economic returns. What if making more products drastically reduced costs? What if as prices dropped people bought many more products? This was the economic equivalent of perpetual motion, and it seemed fantasy at the time.

The idea has come up several times since, but the researchers abandoned work for two reasons. First, the equations that predicted such behavior were very complex and almost impossible to solve without supercomputers, which had not yet been invented. Second, they could find no real world markets that acted this way. The ideas seemed silly. Certainly, the farms and factories of the day did not exhibit such behavior. Or did they?

Almost a century before Henry Ford had done much the same thing with cars, but the lesson was forgotten in the chaos of the depression. Auto companies had stopped bothering with trying to force what they called "economy of scale," because when the world economy collapsed there was no money to buy cars, however cheap they might be. The world wasn't ready for the Information Age in Henry Ford's day. 1930's technologies for building cars didn't allow steep learning curves. Ford's attempts to cut costs were increasingly abusive of workers. In those days, to dig a bigger ditch, management said, "shovel faster." Too many managers still use such methods.

The General Motors model, mass marketing of merely adequate products into pent up demand, dominated in the years after World War II. The learning curve for cars flattened and the demand curve collapsed until the Japanese recreated the phenomena by using quality, modern technology, and lean manufacturing to improve products. For the rare product-market areas lucky enough to possess both steep learning and demand curves, dynamic prosperity can result. Hicks' speculation is no longer conjecture, but proven fact. Under the right conditions, a semblance of economic perpetual motion is possible.

However, in the 1980s, radical new markets were invented. If you could cut the price of a computer by a hundredfold, sales might go up a thousandfold. This happened, not just with PCs, but also with VCRs, video cameras, FAX machines, and CD players. As innovations diffuse in large quantities, there is a self-reinforcing cycle where the positive impact of this diffusion leads to increased demand for the product and the growth of complementary products.

Economists have used the phrase "increasing returns" to describe this phenomenon.[11] Knowledge-based industries, such as computers, pharmaceuticals, missiles, aircraft, automobiles, software, telecommunications, and fiber optics, exhibit this pattern of behavior. Since knowledge is a creation of the mind, and hence limitless, there are no bounds to growth.

While these industries require large capital investments in research and development, and are generally difficult to design and manufacture, once sales begin, incremental production is relatively cheap. Increased production brings additional benefits: producing more units means gaining more exposure to manufacturing and achieving a better understanding of how to produce additional units even more cheaply. Moreover, experience gained with the product incorporating similar or related technologies. Japan leveraged its initial investment in building precision instruments into capacity for building consumer electronic products, then integrated circuits. Korea followed the same example.

In short, new products do not always mature and die in the conventional sense. Instead, they set off new patterns of renewal and growth based on the amplification of customer demand. This is illustrated in Exhibit 6-2. Growth accelerates and intensifies as "lock-in" — the third engine of prosperity takes place.

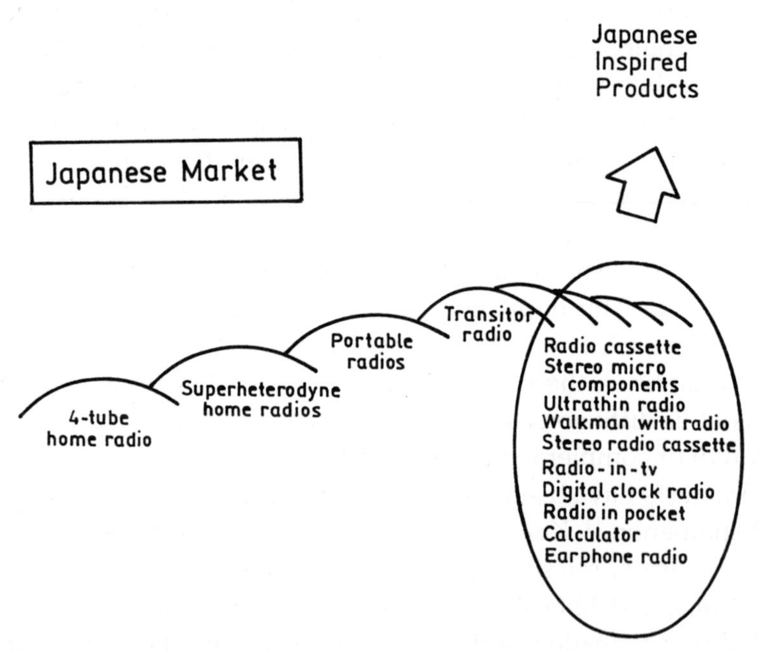

ADAPTION SOURCE: Adopted from Kenichi Ohmae, *The Mind of the Strategist: The Art of Japanese Business*. New York: McGraw-Hill, 1982.

Exhibit 6-2: Illustrating the Growth of Complementary Products (and Increasing Returns) Using the Product Development of the Radio.

Engine #3: Technology Generators, Bandwagons, and Lock-in Processes

In industries where standards are important, the ability of a firm to successfully establish its technology as the standard becomes the basis of its long-term success and viability.[12] The diffusion of such technological standards or "drivers" increases the innovative uses of the product, as well as the development of complementary products. One example was Mostek's 64k DRAM that became an industry standard. In obtaining valuable experience in manufacturing random access memory chips, for example, firms can effectively produce the next series of complex circuitry (256k DRAM).

Another example is Intel's successive introductions of the 286, 386, 486, and Pentium microprocessors that has defined the essential design standard for many generations of personal computers. Intel's microprocessors were commercially more profitable than DRAMs, because Intel owned more of their intellectual content. Even with strong patents and intellectual property protection, and despite aggressive assertion of ownership rights, some cloning of Intel's processors has occurred.

According to the work of Brian Arthur, in contestable industries where two or more incompatible increasing return technologies compete, small changes in initial conditions, whether due to chance or strategy, can lead to one technology gaining the lead to eventually lock in the market and become the *de facto* industry standards, with other competing technologies being effectively locked out.[13] This may happen even when the accepted standard is clearly inferior to other designs.

The classic example of the market locking in to an inferior technology is the QWERTY format for typewriters, now extended to computer keyboards.[14] Originally developed by trial and error in the 1860s to compensate for a design deficiency in typewriters, it became a lock-in standard even when better engineering attenuated this initial problem. Despite superior keyboard formats that could have increased typing speeds considerably, the QWERTY format prevailed. This was sustained when first touch-typists, who had trained in the QWERTY format, developed this preference. Eventually, training on the QWERTY format was so widely adopted that it became institutionalized.

Many of today's existing designs have resulted from lock-ins made several decades ago. Consider the case of the Florence Cathedral Clock for which its hands moved in a "counterclockwise" direction around the 24-hour dial when first designed by Paolo Uccello in 1443. At the time a convention for clockfaces had not yet emerged. Competing designs were subject to increasing returns: the more clockfaces of one kind were built, the more people became used to them. Hence. it was more likely that future clockfaces would be of the same kind. After 1550, "clockwise" designs displaying only 12 hours had crowded out other designs. This is reproduced and illustrated in Exhibit 6-3. W. Brian Arthur, one of the leading advocates of the theory of positive returns to scale, argues that chance events couple with positive feedback, rather than technological superiority, often determine economic developments.[15]

Lock-ins are inextricably tied to technology generators and bandwagons. Once "locked-in", technologies tend to be difficult to change in the absence of another revolutionary breakthrough. Competitive imitation also plays a role in sustaining locking-in processes. Suppose that firms enter an industry one by one and choose their locations so as to maximize profit. The geographical preference of each firm varies; chance determines the preference of the next firm to enter the industry. Also suppose, that the firm's profits increase if they are near other firms (their suppliers or buyers). The first firm to enter the industry picks a location based purely on geographical preference. The second firm decides to enter the market based on preference modified by anticipated benefits of being close to the first firm. The third firm is influenced by the first two, and so on.

If some location by good fortune attracts more firms than others in the early stages of this evolution, the probability that it will attract more firms increases. Industrial concentration becomes self-reinforcing. This also helps explain the success of various regional clusters, such as

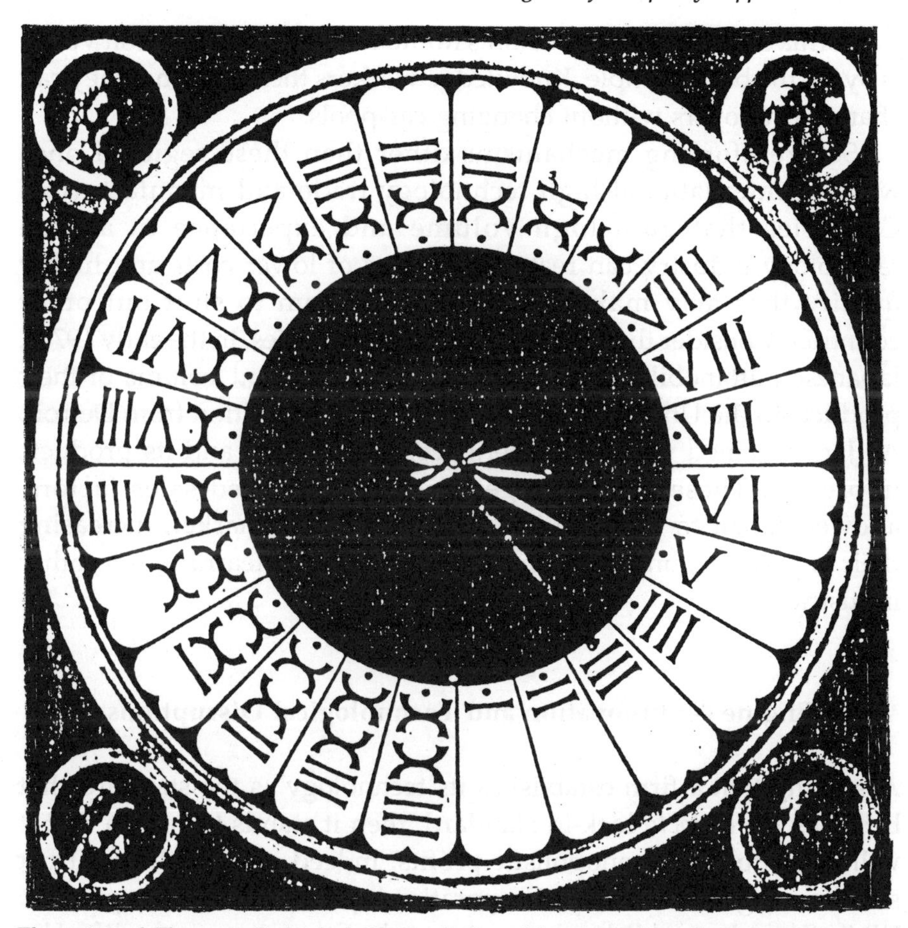

The original Florence Cathedral Clock, designed by Paolo Uccello in 1443, featured hands that moved in a "counterclockwise" direction around the 24-hour dial. In 1550, the clockwise designs had replaced the original design and become the "lock-in" for future designs.

SOURCE: Adapted from W. B. Arthur, "Positive feedbacks in the economy," Scientific America, February 1990: 94. Original illustration by Casa Editrice Giusti di Becocci, Firenze, Italy.

Exhibit 6-3: Illustrating a Lock-in Phenemona With the Florence Cathedral Clock.

California's Silicon Valley, Boston's Route 128, and others. Silicon Valley is by far the best example. Knowledge workers there joke that they can change employers without changing car-pools.

Self-reinforcing mechanisms other than these regional ones work in international high technology trade and manufacturing. Countries that gain high volume and experience in a high technology industry can reap advantages of lower costs and higher quality that may make it possible for them to shut out other countries. Consider the case of Japanese automobiles. In the early 1970s, Japanese automobile firms started to sell significant volume of their products to the United States, without much opposition from Detroit. As Japan gained valuable experience, its costs fell and its products improved. These factors, together with improved sales factors, allowed Japan to increase its share of the U.S. market, increasing their experience, and allowing them to leverage learning into other industries.

Engine #4: Innovation and Technological Disruptions

Assuming that a firm establishes its technology as the standard that becomes the *de facto* lock-in standard, then it stands to reason that it will begin to reap monopolistic profits. Indeed, part of the antitrust suit levied against Microsoft Corporation is premised on this argument. While firms do establish a competitive advantage — as exemplified by Intel and Microsoft — such a lead is not guaranteed.

An informative study by Clayton Christensen examined how industry stalwarts, such as Sears, IBM, DEC, Data General, and others, were blindsided by erstwhile competitors after they had built what appeared to be insurmountable advantages, partly from their abilities to lock-in their technologies.[16] Remarkably, firms failed not because they did anything wrong, but even when they did "everything" right, including listening to their customers, and investing in new technologies that were oriented at serving their customer base.

To understand this dilemma, Christensen contrasts "sustainable technologies" from "disruptive" technologies. Sustaining technologies continue the rate of improvement of established products. For example, in the disk drive industry, established companies sought to improve the drives they were making in order to offer higher capacity at a lower cost per megabyte. Disruptive technologies, on the other hand, typically results in less product improvement but offer features that new customers value. These are typically smaller, simpler, and more convenient to use.

While established leaders in the disk drive industry were increasing the capacity for the 14-inch drives to power mainframes, new companies such as Shugart, Micropolis, Quantum, and Priam developed 8-inch drives that offered less capacity. Nevertheless, these disruptive drives were eventually used in new products — minicomputers made by Wang, DEC, and Hewlett Packard. Ultimately, every 14-inch drive maker was driven from the industry.

The makers of the 14-inch disk drives failed because they listened, perhaps too intently, to their established mainframe customers, who said that they did not want a smaller drive. Yet, technologies can progress faster than the demands of the market. In efforts to satisfy their customers, firms can "overshoot" the market, giving customers more than they can handle, when, in fact, disruptive technologies that offer less may be sufficient. Because of the rapid pace of technological developments, disruptive technologies soon catch up with customers' needs. Meanwhile, firms that did introduce these disruptive technologies may have already established their competitive advantage over more entrenched rivals. Previously established firms find no recourse other than to belatedly jump onto the bandwagon to defend their customer base, starting yet another cycle of competitive maneuverings. The implication is that companies must develop new markets that value disruptive products, rather than trying to adapt the technology to mainstream markets. In Exhibit 6-4, we illustrate changes in market leadership of selected industries over a number of decades.

INDUSTRY	1960	1970	1980	1990
AEROSPACE	McDonnel Douglas	Norris Industries	Lockheed	Teledyne
AIRLINES	American	Delta Air	Emer/Delta	UAL
APPLIANCES	RCA	Magnovox	Maytag	Hormon Int.
ALUMINUM	Kaiser	Kaiser	Kaiser	Alcan Aluminum
AUTOMOTIVE	General Motors	General Motors	Subaru	Ford
AUTOPARTS	Libbey-Owens-Ford	Champion	Snap-On-Tools	Bandag
CAPITAL GOODS	Caterpillar	Caterpillar	Caterpillar	Tecumseh
CHEMICALS	Du Pont	Eastman Kodak	Dow Chemnicals	Union Carbine
DRUGS	Smith-Kline-French	Smith-Kine-French	Smith-Kline-French	American Home Products
DEPT, STORES	J. C. Penney	Zayre	Dayton Hudson	Dillard
OIL	Shell	Ashland	International Oil	Exxon
FOREST PRODUCTS	Scott Paper	Crown Cork & Seal	Willamette Ind.	WTD Industries
STEEL	US Steel	Kaiser	ARMCO	Allegheny
RUBBER	Goodyear	Goodyear	General Tire & Rubber	General Tire & Rubber
TEXTILES	Burlington	Melville Shoe	Cone Mills	Delta Woolside

SOURCE: Adopted from various issues of Fortune magazines: January 18, 1960; January 12, 1970; January 22, 1980; and January 14, 1990.

Exhibit 6-4: Changing of the Guard: Market Leaders Over Three Decades.

Engine #5: Outsourcing and the Unraveling of a Firm's Value Chain

How do firms compete in an era characterized by the continuous disruption of successful products? The value of a firm has been described in terms of its underlying value chain, or its discrete set of activities relating to its infrastructure, inbound, throughput, and outbound activities. We argue that change emanates from changing economies that tend to unbundle a firm's value chain. This case is most pronounced in industries, such as semiconductors, telecommunications, or health care services, that are characterized by both consolidating and fragmenting structures.

In semiconductors, for example, traditional designs called for the entire process — from design conception to volume production — to be realized within single firms, i.e., fully captive, merchant firms. This allowed them to retain control over the entire value chain, with the end in view of generating sufficient capital to cover entire process, i.e., the costs of design, development, and manufacturing — as well as generating margins for new product development. As design costs for customized chips become more costly, and as manufacturing became more efficient, new entrants attempted to lower the costs of custom circuits by unbundling the entire process. This was eventually achieved by transferring, or diffusing semiconductor design technology to systems producers. In this way, the costs of design and development became reduced through automation and specialized. At the same time, production technology was offered as a foundry service at the open market. That is, completed masks were farmed back to new merchants for prototyping and final production in foundries. This unbundling effectively altered the traditional economics of semiconductor manufacturing and allowed various competitors to amortize the high capital costs of design and development through sharing. Production costs were also shared across a large number of designs, thereby incurring more economies. See Exhibit 6-5 for this illustration.

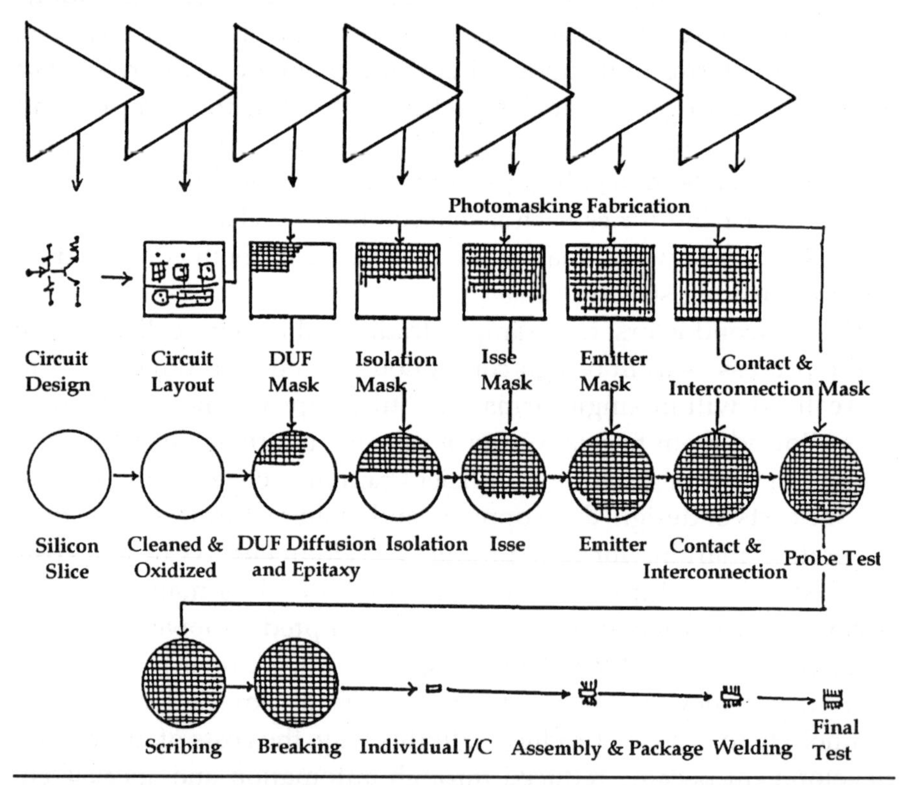

Traditional competition focused on the tight bundling or interconnections between manufacturng activities the defined the firm's value chain. Advantages derived from cost and quality were centered at the final assembly stage of the product

Exhibit 6-5: The unbundling of semiconductor manufacturing processes: traditional and contemporary views.

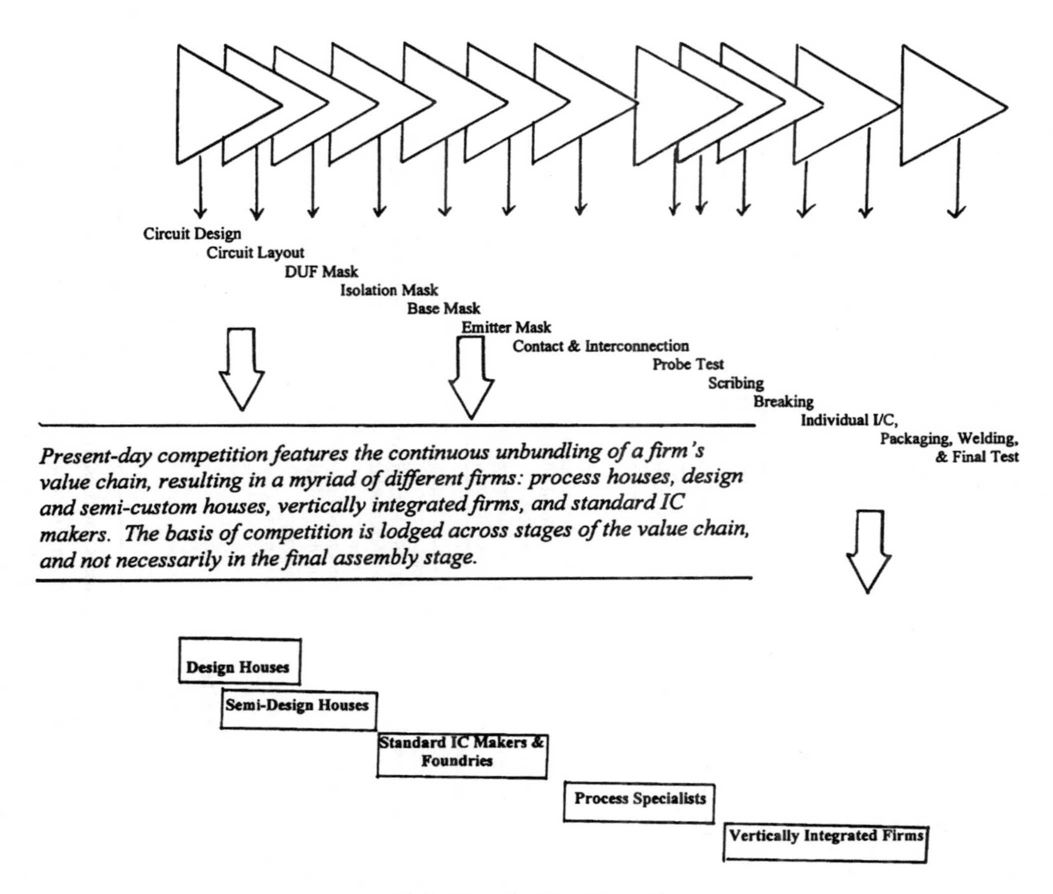

Exhibit 6-5: *(Continued)*

Standard IC Makers

Advanced Micro Devices
Cherry Semiconductor
Intel
Eurosil Electronic
Exar
Fairchild Semiconductor
General Instrument
Inmos
International Rectifier
Matra-Harris
Micron Technology
Monolithic Memories
Motorola Semiconductor Products Sector
National Semiconductor
SGS Microelettronica
Siliconix
Sprague
Standard Microsystems
Texas Instruments
Western Digital
Zilog

Vertically Integrated Companies

AT&T
Ford Microelectronics
GE/RCA/Intersil
GM/Hughes Electronics
Gould
GTE Microcircuits
Harris
Honeywell
ITT Semiconductor
Mitel
NCR
Philips/Signetics
Plessey
Raytheon
Rockwell International
Siemens
Telefunken Electronic
Thompson Components-Mostek
TRW
All major Japanese and Korean semiconductor Producers

Design Houses

Digital

Altera
Brooktree
Dallas Semiconductor
Lattice Semiconductor
Chips & Technologies
Cirrus Logic
Faraday
Integrated Device Technology
Logic Devices
MOS Electronics
Vitelic
Weitek
Xillinx
Zoran

Linear

Anadigics
Analog Devices
Burr-Brown
Crystal Semiconductor
Linear Technology
Maxim
Micro Linear
Precision Monolithics
Silicon Systems

Process Specialists

Bipolar Integrated Technology
Catalyst Semiconductor
Cypress Semiconductor
Exel Microelectronics
Gazelle Microcircutis
GigaBit Logic
Hypres
Inova Microelectronics
International Microelectronic Products
Micro Power Systems
Mosalc Systems
Orbit Semiconductor
Performance Semiconductor
Seeq Technology
Sierra Semiconductor
TriQuint
Vitesse Electronics
Xicor

Semicustom Houses

Actel
Applied Microcircuits
California Devices
Ferranti
Integrated Logic Systems
International Microcircuits
LSI Logic
Mican Associates
Micro Linear
Solid State Scientific
VLSI Technology
Waferscale
Zymos

SOURCE: Adopted from various sources, including Stan Avgarten. *State of the Art: A Photographic History of the Integrated Circuit*, New Haven, New York: Ticknor and Fields, 1983; T. R. Reid, *The Chip*, New York, New York: Simon and Schuster, 1984; and Rodnay Zaks, *From Chips to Systems*. Berkeley, California: Sybex Incorporated, 1981.

Exhibit 6-5: ***(Continued)***

Michael Borrus and John Zysman argue that competition has shifted from final assembly and vertical control of markets by final assemblers.[17] They define this shift as a struggle over setting and evolving *de facto* product standards in the market, with market power lodged throughout the value chain, including product architectures, components, and software. In their view, the latter become separate and critical competitive markets.

Historically, the electronics era was dominated by assemblers, i.e., systems producers who designed, marketed, and assembled the final product.[18] Early American industry, with dominating icons like GE, RCA, General Motors, and IBM, prospered with traditional advantages associated with scale economies, vertical integration, and mass production.[19] Starting in the 1960s, however, in their attempt to emulate American producers, Japanese enterprises such as Matsushita and Hitachi began to dislodge established American positions in the consumer electronics industries by applying the lean production techniques and improved production designs. Their competitive strengths were abilities to manufacture high quality at consumer price points with some degree of product variety.[20]

By as early as the 1980s, essentially all electronics product makers were dominated by large scale producers such as IBM, Siemens, Matsushita, and NEC, and Toshiba, that produced fully proprietary systems, whereby key product standards were either fully "closed" or "fully open." A fully open standard is one in which the technical information necessary to implement the standard is in the public domain. With relevant technical information in the public domain, products like TVs and radios built to such standards became commodities in which scale, quality, and cost were defining features of competition in highly contestable markets.[21]

By contrast, telecommunications and computer firms built to "closed" standards in which the relevant information was owned as intellectual property and not made available to anyone other than through legally permissible reverse engineering.[22] Vertical integration and manufacturing were essential. Once established, competition centered on a growing installed base of customers who were locked-in to a firm's product line. Lock-in was possible because of the high costs of switching between closed systems. With both closed and open systems, vertical control over technologies and manufacturing was the key to market success. It was necessary to capture closed systems rents

Exhibit 6-6

What explains the patterns of success and failure in key industries? Why do some industries undergo recurring patterns of growth and rejuvenation? Can new technology reinvigorate old industries, such as textile, merchandising, steel, and automobiles?

The Engines of Prosperity	*What It Means*	*How It Is Reflected*
Steep Learning Curves	Efficiency arises with increased production or service transactions	Learning curves, i.e., in semiconductors, every doubling of production results in about a 30% reduction in costs.
Demand Amplification	High elasticity, or that a drop in prices results in a highly dispropor-tionate increase in sales of both the product and complementary products	Pricing strategies of new software
Technology Generators/ Lock-in	An industry standard becomes the "driver" of future generations of product, i.e., "lock-in" effect	Mostek 64k DRAM; Bandwagon, Intel's 286, 386, 486 microprocessors
Innovations/Disruptions	Massive diffusion of applications offer similar, complementary, or disruptive products	MS-DOS; Windows New entrants
Outsourcing/Value Chains	Continuous unraveling of firms' value chains	Intel vs. Netscape; Intel vs. Java

and lock in customers into proprietary standards, or, in the case of open systems, to compete on implementation, quality, and price.

A summary of these arguments are presented in Exhibit 6-6. Understanding the dynamics of how these engines work together will provide new insights on how to manage this environment. The rise and fall of strategic management models reflect their ability to come to grips with changes in the business environment. These changes, in turn, may explain why competitive advantages in today's age are so difficult to sustain.

The Engines in Context: The VCR Industry

The videocassette recorder industry is undoubtedly the exemplar example of how the Engines of Prosperity can sustain an industry.[23] Back in 1951, David Sarnoff, chairman of RCA, challenged his staff to develop a television picture recorder that would record the video signals of television. RCA was beaten to the marketplace by upstart Ampex Corporation, a small California engineering firm that introduced the first commercially viable videotape recorder (VTR) in 1956.

The VTR units were very expensive, hundreds of thousands of dollars, and were sold for commercial broadcast applications. These markets were limited and small in size. Competition shifted to developing the video recorder for mass-market home use — a feat claimed by Sony in the 1980s. Currently, the consumer market is dominated by the Japanese and the Koreans, with Europeans having minor presence, and the United States, who invented the technology, with hardly any representation at all.

Ampex's experimental model (VR 1000) was an initial success and the company aggressively pursued its installed base in the broadcast market. The technology was also patented such that no other company was able to legally manufacture a VTR without a license from Ampex. While popular, the original four-head design was difficult to operate and maintain. Seeking to improve the product, and seemingly confident of its strong market position, Ampex entered into cross-licensing agreements with RCA to achieve color recording. In 1959, Ampex also exchanged its technology for Sony's help in developing a VTR with transistorized circuits. However, this partnership ended after two years over a disagreement about royalty payments related to Ampex patents.

In the bull market of the 1960s, Ampex was one of the glamour stocks of Wall Street. However, Ampex was also vulnerable to tight margins with emerging competition from both audio and video markets. Its response to the needs of the mass market was the Instavideo, a lightweight, easy to use machine with playback and recording capabilities. Initially priced at $1,000, it was an instant success but the announcement also harbored Ampex's weakness: it did not have the requisite skills or facilities to manufacture a technologically sophisticated product in large volumes at low cost. In view of its cash-constrained position, it decided to use the manufacturing capabilities of Toamco, Ampex's joint venture with Japan's Toshiba. Beset with financial difficulties, Ampex management restructured, closed plants, discontinued some businesses, and terminated the Instavideo project.

RCA, Ampex's archrival, opted for a more complex videoplayer, which it introduced under the name Selectavision in 1969. This utilized lasers and holographs for the first time in consumer application. However, early technical problems prompted RCA to pursue other technologies, notably the videodisk, with the purpose of providing a reliable machine with high quality picture for less than $500. Even so, the product lacked recording capabilities, and sales proved quite disappointing. In the spring of 1984, RCA announced that it would discontinue production of the Selectavision VideoDisc. Relatedly, CBS also tried its hand at the videorecording market, but could not deliver its product (Electronic Video Recording System) on time and eventually withdrew from the market. The only other U.S. company was Cartridge Television that encountered problems with hardware, tape quality, and software availability, forcing the company to declare bankruptcy in 1973.

While U.S. companies struggled through the development of videorecording technology, Japanese companies, by contrast, took initial failures in stride. Japan's broadcast network, NHK, imported an

Ampex VR-1000 video recorder in 1958 and placed the machine on display to encourage Japanese electronics companies to build their own VCR. Sony, along with the NHK lab, built a replica that was bulky and expensive. However, the experience convinced Sony that a consumer videorecorder could be a logical extension of the company's product line. Creating this new market would take over two decades.

In the late 1950s, Sony had maintained a project team to work on a home videorecorder. The company's first VTR was designed for educational and industrial use, and these were used mainly by American Airlines. In 1971, Sony introduced the first videocassette recorder (VCR) for the consumer market. Still, the "U-Matic" machines and cassettes were too big and expensive and were relegated to industrial use. This experience, however, kept motivating Sony to design a relatively inexpensive machine for consumer use.

With the introduction of the EVR system and the Selectavision, Sony approached JVC and Matsushita — two of its biggest competitors — about developing a Japanese standard around a new Sony technology which would reduce machine size. Even with the three companies agreeing to share existing technology, the competitors would only accept Sony's basic U-matic format, and Sony and JVC refused to cooperate or compromise on technology for smaller machines.

Despite losses, both JVC and Matsushita continuously maintained VTR development efforts targeted at the consumer. At JVC, the manager instructed his team of marketing, production, and design specialists not to ask what was technologically possible, but to "determine what consumers wanted in a home VCR and then develop the technology to meet those requirements."

In the early 1970s, Sony identified 10 different ways of building a home VCR and created research teams to explore all these alternatives. By 1974, it decided on a prototype called the Betamax, and it invested in new plant facilities for mass production. In order to make the machine more compact, Sony restricted the length of its recording capability, since it believed that an hour of recording time was sufficient.

Sony asked Matsushita and JVC to adopt the Beta format, but both rejected, citing the one-hour recording time as a major limitation. JVC was also convinced of the superiority of its VHS format, designed to deliver up to three hours of taping. After the Betamax introduction, Hitachi approached Sony about licensing the technology, but this request was refused on grounds that the Beta technology was still not perfected.

Taking a different track, JVC sought to form alliances with Japanese companies that would agree to accept the VHS as the standard format before even shipping its first product. A tentative alliance was developed with Matsushita, Hitachi, Mitsubishi, Sharp, Sanyo, and Toshiba. Two warring groups, Sony versus JVC, emerged, with Sanyo and Toshiba eventually joining the Beta fold. By the end of 1977, almost 80% of VCRs sold in the two biggest discount districts of Japan were Beta. Even so, by the end of 1978, Matsushita replaced Sony as the share leader. By 1988, VHS had close to 95% of world sales, leading Sony to start manufacturing a line of low-end VHS machines alongside its high end Betamax machines.

Rather than producing VCRs, American firms opted to market the machines through OEM agreements with the Japanese. Zenith received VCRs from Sony, while Sanyo supplied Beta-format machines to Sears. Magnavox and GTE Sylvania agreed to sell the VHS format supplied by Matsushita. RCA reached agreement to supply Matsushita VHS machines capable of recording for four hours. U.S. market penetration for VCRs rose from less than 2% in 1979 to more than 50% in 1987. The increased sophistication of VCR machines along with the growing availability of prerecorded tapes stimulated sales further.

In the early 1980s, the VHS format made up about two-thirds of the VCRs sold in Europe, while Beta accounted for almost a quarter of the market. In contrast, Philip's entry, the V-2000, accounted for a little more than 10%. The Europeans then resorted to protectionist measures to halt the entry of Japanese products, starting auspiciously with an announcement that all imports of VCRs could only be cleared through Poitiers. A few weeks later, Philips and its VCR partners Grundig filed antidumping suits against the Japanese VCR exporters — the largest antidumping suit in the history of Japanese-EC trade.

Recent history of the VCR features the entry of Korean consumer electronics companies into the VCR market. They had signed agreements licensing VCR technology from a number of Japanese companies. The Koreans entered the U.S. market in 1985 with relatively inexpensive, promotional brands that were usually supplied on an OEM basis to mass merchandisers. The median price of a Korean VCR was approximately 30% below that of a Japanese VCR . As product quality improved in 1986, Korea became the second largest VCR exporter in the world with a 9% global share. More than half of Korean sales went to the United States.

An interpretation of the development of the VCR provides context to the Engines of Prosperity discussed in the first part of the chapter. The dramatic growth of the VCR is founded on two interrelated concepts: falling costs and market acceptance. As costs fall, prices fall accordingly, and products become more accessible to mass consumers. However, for costs to fall, it is important that scale and scope economies are realized (*Engine #1*). In our description, the Japanese were able to achieve this, while American producers, limited by management concerns and their financial institutions, were not.

U.S. firms, needing near term quarterly profits, simply could not compete. Therefore, the only U.S. participation was in reselling foreign made VCRs under local brand names. Since this add less value than the full value manufacturer of goods, most of these firms were purchased in the end.

In the beginning, the market was led by Sony's Betamax format, with VHS as the follower. Because the market was unstable, market shares for these formats fluctuated based on external circumstances, luck, corporate maneuvering, and responses to consumers' preferences. Eventually, the firms favoring VHS made better use of learning curves and demand amplification. Increasing returns on early gains led to the dominance of this format.

Moreover, the acceptance of the VCR was aided by amplified demand for supplementary products (*Engine #2*). While this point will be more apparent in the case of semiconductors, it is clear that market acceptance for the VCR came with the availability of prerecorded tapes, in addition to the VCR's core function of recording television broadcast services. Moreover, the popularity of X-rated prerecorded tapes altered lifestyles and movie-going behaviors, and these served to boost initial VCR sales. Currently, the development of the videocameras, camcorders, coupled with widening popularity of homemade movies, have further sustained the commercial longevity of VCR machines. Other supplementary products include 8 mm camcorders, the 3-inch flat screen, pocketbook sized 8 mm VCRs, and bundled VCRs, TVs, and audio equipment.

Despite Sony's introduction of the Beta technology, it could not command sufficient market share to keep out crowding technologies such as the VHS. Once "locked-in," however, the VHS stayed ahead and increased its lead. Soon, despite its utmost efforts, a superb competitor (Sony) with a clearly superior technology (Beta) was totally "locked-out." The only hope Sony has to regain leadership is to somehow create a new equilibrium point, perhaps by exploiting new technology (*Engine* #3).

To sustain commercial development, products have come by more cheaply, thus amplifying the investment-market acceptance cycle. This is borne out with the early experience of the Japanese, and with the more recent experience of the Koreans. The latter have competed on the basis of a no-frills, cheap but reliable image. As VCR prices continue to drop, then markets will broadened to even more segments, increasing product acceptance (*Engine #4*). By applying lean manufacturing techniques, Koreans have been able to dismantle barriers by established leaders and compete effectively in one end of the product spectrum (*Engine #5*). In contrast, the failure of the Europeans to sustain momentum beyond their borders is due to their relatively higher costs of production.

Already, the Koreans have become a powerful force in global markets. Lurking in the background are lower wage countries, such as Malaysia, the Philippines, Indonesia, Thailand, and, of course, China. Should production be extended to these countries, either through OEM arrangements or through their own indigenous efforts, then VCRs can reach even higher frontiers. While the Engines of Prosperity as we have defined them offer insights in interpreting developments in the videorecording market, these are even more salient in the case of semiconductors, often referred to as the "food chain" of commercialization.

The Engines in Context: Semiconductors

Semiconductors refer to materials, notably silicon, that function as conductors, permitting electrical current to flow, or as insulators that cut off the flow under particular circumstances.[24] Semiconductors, when used to control the input of electric power, can transmit coded signals for particular uses (i.e., computer, satellite, and consumer appliances).

The invention of the semiconductor is traced to the work of two men: Jack Kilby and Robert Noyce. Acting independently, they invented the integrated circuit. On September 12, 1958, a group of Texas Instruments executives gathered in Kilby's residence to see if this tiny oscillator-on-a-chip, half an inch long and narrower than a toothpick was finally ready. The integrated circuit, the answer to the tyranny of numbers, had worked. The men in the room looked at the sine wave again. Then everyone broke into broad smiles. A new era of electronics had been born."[25]

At about the same time, Robert Noyce worked out the idea, dubbed the "Monolithic Idea", in words quite similar to those Jack Kilby had entered in his notebook six months before, "...it would be desirable to make multiple devices on a single piece of silicon, in order to be able to make interconnections between devices as part of the manufacturing process, and thus reduce size, weight, etc. as well as cost per active element."[26]

At the time, they might not have anticipated the full impact of what they had invented: in the integrated circuit, they had ushered in the era of microelectronics, and the basis for what we presently refer to as the "high technology" industry. What is particularly striking about the invention of the integrated circuit, that contrasts with the above inventions, is the pervasiveness of its effects in a relatively short period. As semiconductor technology developed, the miniaturization of electronics created new markets and revolutionized applications data computing, communications, test and measurements, and industrial equipment.

New consumer products such as the personal computer, digital watches, sophisticated alarms and sensor devices, programmed sewing machines, and automotive controls have evolved into fruition with the new technology. As the miniaturization of electronics continues, even newer industries like the automated office, financial services, transportation, and artificial intelligence will become stronger components of our industry. The industry represents how our five Engines of Prosperity come into play into describing the dramatic rise and significance of the integrated circuit. It is no wonder that one high technology executive once referred to the chip as "the crude oil of the 1990's."

The world semiconductor industry was once completely dominated by U.S. firms. Earlier inventions of the integrated circuit and subsequent improvements were based on a massive influx of defense dollars, notably the Minuteman II missile program. Antitrust regulations at the time excluded American Telephone and Telegraph (i.e., Western Electric) and International Business Machines (IBM) from entering the industry, creating opportunities for smaller merchant firms for which a number of major innovations and process modifications originated. Even with the subsequent entry of IBM and AT&T as major captive producers, the structure of the semiconductor industry was a viable one, with small-scale producers serving "niche" markets that were not as successfully serviced by the larger merchant firms.[27]

Since the invention of the transistor in 1947, the U.S. has enjoyed economic and scientific supremacy in high technology, with modest competition from Great Britain. The success of U.S. firms as innovators in the field was facilitated by early Pentagon funding in semiconductors. The climate for innovation was enhanced by the unique structure of a "Silicon Valley" that combined the benefits of venture capital funding, University research, and a cluster of budding entrepreneurs into one area.[28] Moreover, the implosion of growth markets (computers, software, CAM/CAD, etc.) arising from new innovations in microelectronics was easily accommodated by the large size of the U.S. domestic market, paving the way for even more innovations.

In the past, Japan had relied on both the U.S. and Europe for key innovations which they secured through licensing and cross-licensing agreements. Through superior production techniques, ingenious "reverse engineering" capabilities, and institutional support from the Japanese government and ministries.[29] Japanese firms have excelled in driving down the costs of these products for commercial consumption. Videocassette recorders and compact-discs provide graphic testimony to these accomplishments. Securing licenses from Philips and other companies, the Japanese firms have been able to drive down costs to allow prices acceptable to mass-market consumers — a feat that eluded the inventors of these products.[30]

Playing a "catch-up" role, European firms are presently examining how to work within the constraints imposed by the European Economic Community to become competitive in high technology. While a number of European firms, notably Philips and A.G. Siemens, are not lacking in innovation, as represented by their invention of the videorecorder and the compact-disc, they have not been able to diffuse these products at a competitive level. The limited size of individual European markets, a preference to compete in the U.S. market instead of their own, and myopic managerial strategies developed through the years account for such failures.[31]

Understanding the dynamics of growth in this industry harkens back to what generates growth and development. The growth of semiconductors has been nothing less than spectacular. Gordon Moore, Chairman of Intel Corporation, once predicted that the number of transistors per chip would double approximately every two years. Viewed at the time as a fairly optimistic estimate, this has now been dubbed "Moore's Law" because of its uncanny degree of accuracy. Since the early 1960s, the rate of growth in the complexity of integrated circuits has increased nearly 100% every two-and-a-half years (*Engine #1*).

The growth of semiconductors results from frequent innovations in the base-product that are quickly diffused to other high technology sectors (*Engine #2*). Innovations comprise new product innovations (i.e., transistors to integrated circuits) and process modifications (i.e., CMOS-NMOS-BIPOLAR). An important feature of the semiconductors is their centrality in the electronic information industries. The structure of this overall industry has been likened to an inverted pyramid; at the "upstream" are producers of advanced materials and manufacturing equipment, the "midstream" consists of semiconductor producers, and the "downstream" are numerous end-users of semiconductors.[32] The intensity of the interdependence between these industries distinguishes them from traditional capital-intensive industries. Innovations in semiconductors creates an implosion of applications for the computer, instrumentation, telecommunications, and consumer-related markets.

Diffusion is accelerated by the declining costs of semiconductor devices that accompanies better process improvements and production automation (*Engine #1*).[33] It is also true, however, that advances in semiconductors resulted from developments in one end-user, that is, computer-aided design and computer-aided engineering. This degree of synergy between manufacturers and end-users is what creates opportunities for many high technology sectors. From this context, the weakening of one or more sectors in the pyramid can result in an erosion for the entire spectrum of industries.

Finan and LaMold argue that, prior to 1977, competitive success in semiconductors was based on four factors: consumer acceptance of product design, availability of second-source suppliers, aggressive pricing, and credible delivery acceptance.[34] At the early stage of the product life-cycle, competition focused on product development as several firms would compete to have their product design accepted as the industry standard. Mostek's design of the 16k DRAM is one example of a clear favorite. Once a favorite was selected, other firms would typically enter the race as second-source producers, and competed on the basis of price, marketing and distribution, quality, and reliability (*Engine #3*). Since funds obtained from a mature product were typically invested in a new product, it was important for firms to maintain the cycle of investment and reinvestment.[35]

The Japanese entry into the 64k DRAM market in the late 1970s to the early 1980s changed the competitive rules in the industry (*Engines 4 & 5*). Japanese firms utilized their strengths as low-cost manufacturers of the product to aggressively attack the U.S. market and exploit the financial constraints experienced by U.S. firms during recessionary periods. The success of Japanese semiconductor firms derives from their ability to exploit vulnerabilities of their rivals in managing the economic cycle that is generic to semiconductor-manufacturing. The pattern is depicted using the 64k DRAM as a representative example.

The commodity-like status of the 64K RAM in tandem with lower costs allow the semiconductor firm to hence recoup some of its investment in the early growth stages (*Engine #4*). As with any product, funds from internal sources are critical to financing new products. In this case, forces for innovation arise from large markets for nonstandardized applications (i.e., EPROM and EEPROM) made possible by the design capabilities of VLSI (Very Large Scale Integration). Such applications of advanced circuitry include computer peripherals, analytical testing instruments, laser testing applications, and other software products. The dynamics of the engines are presented in Fig. 2.

The Key Industries as Defined By the Engines of Prosperity: Is There a High Ground?

Understanding the potential of the Engines of Prosperity has become a prerequisite for operating in today's environment. The mere notion of "Engines of Prosperity" (marking crucial technology-market areas where demand and learning curves are steep, and new competitive behaviors emerge) is just such a major paradigm shift. Today most U.S. officials think such a concept is discriminatory and self-serving. They see it as almost an economic equivalent to racism. It is not "fair" for one industry to be favored over another. (Natural laws are indeed unfair, and sometime cruel, as any pilot or engineer can tell you.)

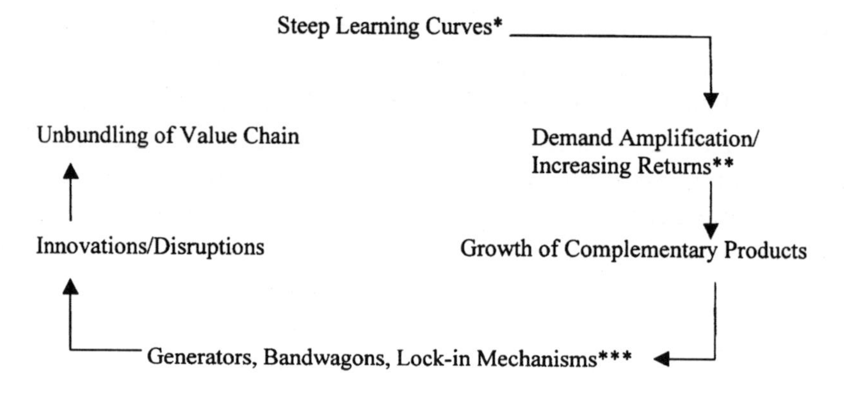

* Steep learning curves lead to significantly higher sales; due to diminishing returns to scale/scope, higher sales at lower prices accelerate growth of complementary products and high margins.

** Lock-in and bandwagon can lead to monopolistic profits (e.g. Microsoft), enhancing the market power of leader (Microsoft).

***Complementary competition in support of the main products would include clone-products. However, others, avoiding a direct competitive collision with leader, form different technological trajectories designed to leapfrog the leader (Netscape, Java).

Fig. 2. The new cycle based on the engines of prosperity.

Those in the U.S. who talk of crucial markets are often labeled as flimflam artists seeking special treatment. Classic Western economic theory says that there are no key markets. Therefore, mainstream economists and trade officials reject such "foolish" notions.

Box 6-1

"Potato chips or semiconductor chips, what is the difference? They are all chips. A hundred dollars of one or a hundred dollars worth of the other, is still a hundred dollars.[36]"

Michael Boskin
Chair
Council of Economic Advisors, circa 1985

"Why do we want a semiconductor industry? What is wrong with dumping?"

Richard Darman
Budget Director (1985)

High technology managers gasp at the above remarks, and dismiss them as the idiosyncratic lunacy of a few people or a single administration. They do this at their peril. Such thinking is very much the mainstream of U.S. finance, politics, and economic policy. This thinking is not administration or even political party specific, but is instead a paradigm that is so deep seated that few in government, education, or economics even question it. As with religious precepts, we educate and train our upcoming leaders to believe the fundamentals. The world's being flat was once a part of navigational, seamanship, and religious doctrine.

Some in the world are using different economic paradigms. This is fact, not opinion. The U.S. thinks that all markets are (or should be) the same, while others — notably the Japanese and Germans — think that certain industries are crucial. The Japanese reduced this to practice first, and certainly they expend much thought on such issues. They have always paid special attention to sunrise and sunset industries. A MITI working group wrote a white paper titled "Japan's Vision for the decade: 1990s." It developed a short list of industries — those with high value added, worthwhile potential size, and *steep learning curves* and steep demand curves (*demand amplification*) — that they deemed crucial to their nation's prosperity. Interestingly, the Germans went through a similar exercise. They independently came up with the same list of key industries.

These lists included market areas like electronics, software, aircraft, and biotechnology. It is the concept that matters, not the lists. The exact lists do not matter for purposes of this book, since they will change from year to year as new technologies and market trends appear. We face a clash of economic paradigms. The U.S. assumes all industries are the same. The Japanese, and others, use the Engines of Prosperity as their model. Europe, through the EEC, is moving from the 1930s vintage laissez-fair model to the new focused economy model.

Clearly, both paradigms can't be right. In this case, we are the ones who are wrong. The test of a paradigm, in the end, is what works. Their model is leading to increased prosperity, while our model is leading to declining prosperity and quality of life. The Engines of Prosperity can be harnessed for national economic advantage.

How do microelectronics, biotechnology, materials science, telecommunications, machine tools and robots, or computers and software differ from other industries? We need to back up a bit and discuss a few concepts that are alien to traditional Western economics. Even today the U.S. government uses "static" economic models that date from the 1930s. How did these come about?

All the markets of the day were slow growth, and based on heavy industry or agriculture. There were no computers or calculators, and economists yearned for simple models that would let them predict trends. They needed the equivalent of Newton's laws, simple formulas that engineers use to predict how objects will act when force is applied.

They developed these models, and, not surprisingly, the test was, "Were they useful for predicting the real world?" The markets of the 1930s were insensitive to external factors. If the price was lower, would you eat more apples, or buy an extra locomotive or automobile? To a first approximation, the answer is "no."

Box 6-2

> **"But in capitalist reality, as distinguished from its textbook picture, it is not (price) competition which counts but the competition from a new commodity, the new technology, the new source of supply, the new type of organization."**
>
> Joseph A. Schumpeter
> (1883–1950)

Models that fit mature, aging economies are dead wrong for today's dynamo markets. Isn't this obvious? Markets driven by the Engines of Prosperity act differently than agriculture or Machine Age heavy industries. Prosperity need not cause inflation. It is well known that the more prosperous the computer or communications industry gets, the lower prices drop.

The Engines of Prosperity can create great wealth. Leading in any of these areas they favor ensures good jobs and a high standard of living. Perhaps more important is that strong positions in any of these can be used to advantage in all the other industries. That complimentary leverage makes the Engines of Prosperity important far beyond the borders of the high technology business world.

While these crucial markets offer the unique possibility of wealth generation that approaches perpetual motion, this is far from assured. Discerning and exploiting these markets is difficult and inchoate art. Most products fail. The winners usually have a new approach, luck, skill, and perseverance on their side, but even the best players strike out frequently.

The whole world is seeking the next PC or VCR. Perhaps it will be HDTV or pen based computing, but no one knows, and no one knows what products will dominate. Japan's first vision for HDTV products was a failure, and most ventures in pen based computing are floundering.

Willy Sutton was once asked why he robbed banks. He answered, "Because that's where the money is." Just as industrial nations once fought for strategic resources and geography, in the Information Age they covet control of the technology-market areas favored by the Engines of Prosperity. These product-market areas are tough and unforgiving, but they hold the keys to future wealth and power.

The markets favored by the Engines of Prosperity are important, and their value extends to things more dear than paychecks and dividends. Simply, if you lead here it almost assures that you will be prosperous. Conversely, companies that lag or blunder are destroyed, stripped of technology, or pushed into less attractive markets.

The law of the learning curve is unforgiving. Firms that fall behind in the dynamo industries rarely survive. Consider that their demise can pull down other industries and perhaps their nations as well. Is it any wonder that U.S. Federal and State bureaucracies substitute short term programs (unemployment, construction, farm or tourism subsidies) for knowledge-dependent family wage jobs? *They don't know what else to do.* In the U.S. government wages are now 110% of private sector wages, but government jobs don't produce products to create national prosperity.

Box 6-3

> **"It is much harder to nullify the results of an economic conquest than those of a military conquest."**
>
> Korekiyo Takahashi, 1936

It is good that cold war has turned to trade war. It is better to decide conflicts and adjust world power by trade than by shooting at each other. Unfortunately, the consequences of losing in trade are at least as bad and perhaps worse than being conquered by military action.

Chapter 7

The Engines of Prosperity: Implications

Key Themes at a Glance

What kind of world will it be where increasing returns to scale, escalating demand, and dense networks of industry families cohabitate? For many, this is not a reassuring thought. Even so, it will most probably spawn an era of investment and growth that we have not witnessed in decades.

Corporate structures are now facing their most radical challenge since the decade of divisional management. Traditional silos, once efficient for solving specialized problems, will soon become a relic of the Machine Age, giving way to virtual, cross-functional teams.

Speed kills. The drive to meet the requirements of fast-changing markets and industry boundaries will place a premium on being first-to-market. But this simply creates an endless cycle of creating new products and services. Distance from the status quo may become the new defensible strategy.

To sustain one's advantage, a firm may have to cannibalize its best products. This is not a failed logic. In hypercompetitive markets, firms need to destroy current advantages in order to move into something else.

> **The advent of virtual cross-functional teams will create new pressures for employee empowerment, shifting flush solutions into mindful action, from process considerations to pockets of experiments. The flesh robots of the Machine Age will give way to a more enlightened workforce.**
>
> **In a world soon dominated by ordered chaos, the emerging management edict will be governed by paradox — when to build, when to destroy, when to harvest, when to renew. Akin to the yin-yang of managerial action, New Think templates force us to confront the open-ended nature of the future of any business.**

In a world increasingly dominated by technological advances and escalating demand, there are important implications for understanding the nature of management and change. Specifically, effective management must be based on assumptions of unpredictability, weak cause-and-effect, nonlinear growth patterns, exponential irregularities, and ill-structured problem solving.[1]

Darwinism is at work and two views battle for the future. One contingent extends the past, ever leaner and meaner. Their assumption is scarcity, and their methods are fear and control. The bleak sweat shops of the early industrial revolution are back. From Oregon to Odessa, workers toil at subsistence wages with little security. Societies decline, managers lose their values, and workers lose hope.

The authors reject this approach. We feel, as with the dinosaurs or Eastern Europe, that it leads to an evolutionary dead end. In a global economy, the model of bosses controlling dumb, cheap, obedient workers leads to endless job cuts and incomes that decline to third-world levels.[2]

A better alternative exists. New technology allows a renaissance, abundance, and freedom for those able to adapt. This will create a huge gap between the old and the new, the trained and the ignorant, the prosperous and the poor. Expect conflict, chaos, and opportunity.

To prepare, managers must develop new mental models and discard old ways of behavior. This approach departs from more traditional and contemporary theories of management and strategy that emphasize predictability, tight coupling between cause and effects, linear models, and systematic problem solving. This chapter presents implications for changing current managerial mindsets.

Shortening the Short Term: The Battle of Life Cycles

It is amusing that when McKinsey and Company generated a simple spread sheet model in the mid-80s, it changed the world.[3] The results were obvious and well known, but the impact of their article was still major.

If market growth was high, but product life was short, then the greatest determinant of profits was getting to market early. In tradeoffs between spending more on engineering or delaying a product introduction, it was better to spend more on engineering. In tradeoffs between manufacturing cost and introduction date, it was better to introduce early, even if manufacturing cost was high.

Fortunately or unfortunately, this simple model is now accepted as gospel, especially in high technology. Trapped on one side by time and on the other by balance sheet management, most firms chose to "go fast" but without spending more. The results have often not been happy, and many firms are in desperate shape. Much like the character in Alice-in-Wonderland who had to run as fast as she could to stay where she was, they are now trapped by the iron laws of the market and of the quarterly report.[4]

The iron law of the market says that if products are similar, customers will select the one available soonest, most proven, most accepted, and cheapest. The iron law of the balance sheet says thou must have increased quarterly earnings. In such a crunch, what can be sacrificed? Obviously innovation takes time and costs money. Therefore, most firms cut their R&D and market creation investments.

The feeding frenzy of the personal computer industry makes Detroit's problems seem trivial. A few years ago there were over two hundred firms selling essentially the same products, and most were losing money. Today the number of vendors is much fewer, but products are even less differentiated, and most firms are still losing money. Cycle times for leading edge PCs are down to six months.

Here is the death spiral: First, a firm slips from the leading edge of innovation. As soon as it stops leading change and becomes a follower, gross margins slip. To preserve profits, it cuts R&D. Then it cut sales costs to next to nothing, it outsources and shares the cheapest standard distribution channels with its competitors. It can't spend enough to stay ahead in technology, and it can't spend enough to know its customers intimately.

Tough competitors, sensing weakness, increase their rates of new product introduction. Andy Grove, for example, chuckles and says, "We'll just go faster and blow these bugs off our windshield."[5] And AMD's (Advanced Memory Devices, a competitor) stock price drops.

Firms trapped in this reactive cycle must run ever faster until they finally collapse. As with any habit, it is very hard and very painful to change this behavior. We advise our clients to invest in innovation, but most of the money in consulting is made by firms who teach simple process improvement and faster cycles without putting much emphasis on value differentiation of products. You know it is hopeless when the CEO of a firm with aged commodity products says, "But that is what we are doing. We are innovating."

The safest place in chaotic, fast change markets, is distant from the status quo. The prosperous firms drive change ruthlessly. They force their competition to react to them. When Intel introduces their next chip, only then can the competition start to clone it. They then fight delaying actions with every weapon at hand, including strategic marketing and legal means, and they have their next product ready for when defenses start to crumble. This is war. As Andy Grove says, "Only the paranoid survive."[6]

This in not to say that fast cycles don't matter for the leaders. They are crucial, but for a different reason. For someone fortunate enough to be in the lead, there is major advantage in delaying the next product as long as possible. You want the project start date to be as late, but the introduction on time. Anyone who has managed high technology product development, knows to avoid the "creeping feature creature."

You freeze the design, and you implement fast. Those who muddle and change almost always fail.

You wait as long as possible so that you can include the newest technology when you start your product development. For example Silicon Graphics introduced a breakthrough desktop computer called Indy in 1993. It was in development for two years, and their CEO said had they started three months sooner they would not have been able to include the digital camera that they deemed a key feature.

That generation of products brought Silicon Graphics a bright flash of fame and fortune. For a time, they were a "celebrity firm" with good earnings and excellent publicity from the use of their workstations in the Jurassic Park movie. Then they faltered, and missed the path to the future.

At this writing, they are a troubled firm trying to recover. The lesson: The Engines of Prosperity are dynamic, not static. You can't stop. There is no defense except motion.

Sustaining Advantage Means Unloading Your Strengths

One other implication from fast cycle competition, is that the leaders must always, repeat always, cannibalize their products.[7] Financially trained managers always have problems in replacing proven high margin products, with unproven low margin products. Reluctance to move almost killed such powerful firms as IBM and Kodak, and it certainly cost the CEOs their jobs.

The cycle for continued product success is shown in Figure 7-1. This was discussed extensively in *High Tech with Low Risk*, but since then other research has become known. [8]The U.S. Air Force asked a question. If you take two "equal" pilots, give them identical fighter aircraft, will one win the battle most of the time? The answer was "yes."

That result was surprising, so they ran simulations and the result was OODA loops. The winning pilot would Observe, Orient, Decide, and Act. Each fighter engagement was a series of cycles, and the pilot

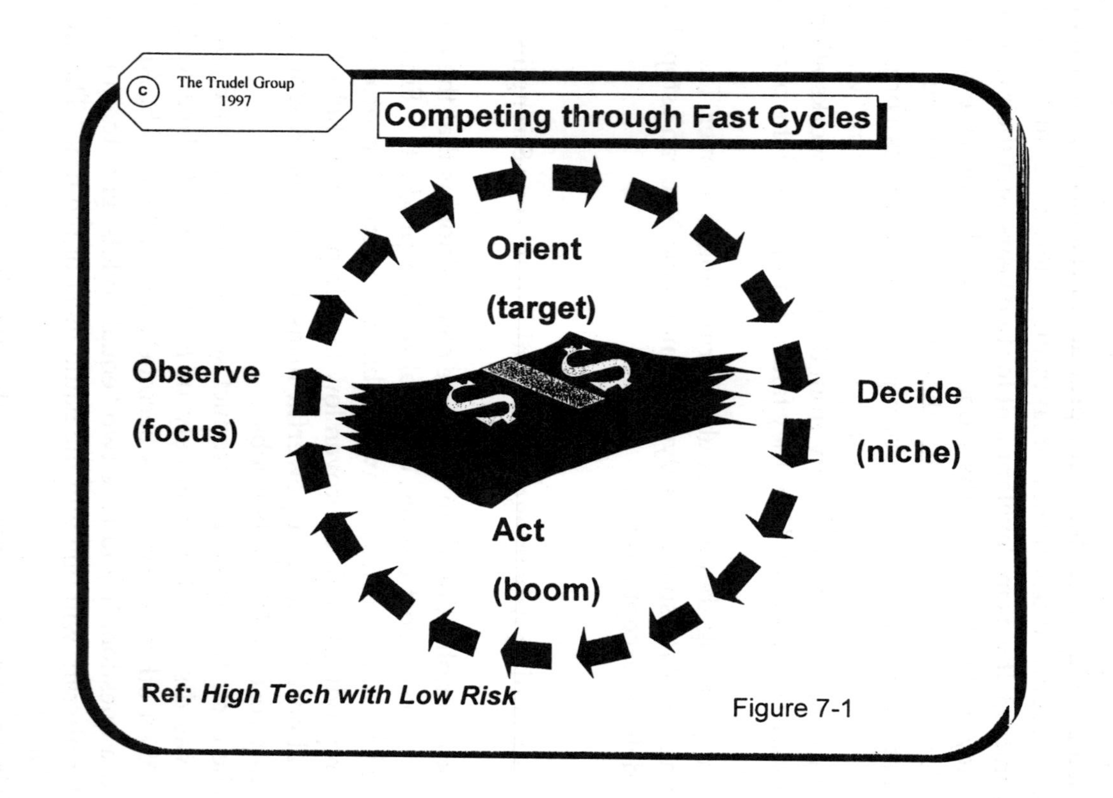
© The Trudel Group 1997
Competing through Fast Cycles
Orient
(target)
Observe
(focus)
Decide
(niche)
Act
(boom)
Ref: High Tech with Low Risk

Figure 7-1

who took the initiative and could go through the OODA loops fastest would slowly force his opponent into ever more disadvantageous positions. The final event, destroying the enemy, was then simple. Competition in the dynamo markets works the same way. It takes continuous cycles of new products to maintain the lead in these markets. Slow cycles are not recommended in either fighter engagements or technology based product competition.

It is an interesting time for Western managers. To win in the market they must innovate aggressively, take risk, spend money, and embrace hard-to-predict strategies that are difficult for competitors to decode. To win in Wall Street they must take little risk, save money, publicize intentions, and produce predictable earnings. The two constraints clash, as do the successful management styles and corporate cultures for each. Since Wall Street often punishes new product investment, it takes a strong CEO to brush off this influence.

Since "slash and burn" is easier to manage, and since Western business serves the stockholders before the stakeholders, it is hard to focus on or adequately fund new products. What results is knock off products and shrinking firms. Some strong firms protect their high margin citadels, as IBM protected its mainframes and Tektronix protected its fast analog oscilloscopes, but technology progress inexorably puts even these last bastions under siege. Once a company chooses "slash and burn" and "clones," it is locked into a behavior pattern that is very hard to break.

Time-based competition is real, but it is also true that speed kills.[9] As the cynics remark, "Death is nature's way of telling you to reassess your strategy." The winners in the dynamo markets must combine speed, innovation, and strategy. Leadership in the high prosperity market requires constant investment in renewal. Learning from the past is not particularly helpful in today's environment, because the next time you do the experiment you might well get a different result. The methods that Apple used with great success to introduce the Mac in 1984 did not work at all well for the Newton in 1993.

Few companies manage to juggle these things well, and fewer still are comfortable with fast cycle chaos. Most of the outside consultants who could be helping find it easier and more profitable to avoid behavioral change issues and stick to improving process.

That's not the answer. Winners today must thrive on uncertainty. The safest way to do this is to create the future yourself. This takes an ability to deliver continuous innovation

Box 7-1

> **"If you are not the lead dog, the view never changes."**
>
> Anonymous Industry Saying

From Separate to Cross-linked Markets

Once one embraces the war metaphor, you start looking at products and markets differently. If you excel at air power, how can you exploit that capability for the benefit of your foot soldiers who may be having tough going?

Cross-linking markets is the key to world power in the 90s.[10] Even though the dynamo markets, the engines of prosperity, are attractive job creators in themselves, they have a vast multiplying effect on one's ability to succeed in other markets. Countries that do this well will dominate world trade and provide prosperity for their citizens. Those that don't, won't.

For an example, let's return to automobiles. Can one include unique technology from the dynamo markets to develop an unfair competitive advantage and raise one's profit margins? Of course, and it is easy. You can use automation and robotics to improve automotive quality and cut manufacturing costs. Toyota does this. You can use materials science to create affordable, lightweight, fuel efficient vehicles. The Lexus bumpers are now molded from an aluminum reinforced elastomer modified thermoplastic, and it gives them competitive advantage. There are many other examples and opportunities.

You can cross-link for advantage within the high prosperity markets themselves. For example, suppose you lead in the machines that fabricate integrated circuits, or in machine tools. That means that your own industry gets to "beta test" the products, and they have a year or two of time advantage before these new tools are available for sale to the general market. A year or so of "free" lead time in a rapid cycle market is an enormous edge.

No one understands the value of cross-linking more than Bill Gates. He has used his control of DOS, the operating system for Personal Computers, to create major business advantage. Just the in-depth foreknowledge of operating system features allows Microsoft a competitive edge in their "applications" business. They have inexorably moved into a number one position in markets like spreadsheets and word processing, squeezing competitors like Lotus and WordPerfect out of the market. One widely reported Microsoft ditty is the saying, "DOS isn't done until Lotus won't run."

Success breeds success, so having expanded to a strong position in applications allowed Microsoft to pressure competitors like Apple, who needed Microsoft's applications available early on their computers, to refrain from competing in both operating systems and application software.[11] In the mid-80s, John Scully was intimidated into licensing Apple's Windows-like operating system to Microsoft and into killing a version of BASIC, MacBASIC, that was superior to the one Gates was selling. This, again, is more like war than genteel business, but the results have made Microsoft one of the most successful companies in the world.

And it doesn't end there. Now that Microsoft has a dominant position in both operating systems and applications for individual ("personal") computing, they are well positioned for a move into multibillion dollar networked computing markets (look out, Novell and UNIX vendors) and the emerging markets like video and multimedia.

You can cross-link for advantage in very mundane industries. For example, if you are in financial trading and you have a unique state-of-the-art computerized communication system that gives you a one minute advantage, you don't make more profit, you make all the profit.

In textiles, high wage nations can't compete with third world nations that pay women at sewing machines a few pennies a day. The solution is to cross link. Clever firms are using high definition television and computer aided design to produce new fashion garments within hours. By the time their products are copied, a few months later, they have moved on to the next fashion generation.

Then there are those boring industries like mass merchandizing. Margins are depressingly low, and competition is ferocious. Until innovators like Wal Mart exploited general aviation, multimedia communications, and computerized inventory control and fleet management to become one of the fastest growing and most profitable companies in the world.

The list is endless, but the point is clear. Those who can control the high ground, the engines of prosperity, will control world wealth and world power in this new information age. The cold war is over, and national prestige and prosperity will be set by trade war. The nations that dominate the dynamo markets, the engines of prosperity, can control competitive advantage in all the industries that matter.

It is the desire for key positions in the future high prosperity markets that drive Japan and force Europe to bury centuries of hostility and band into a common market. It is these pressures that caused four nations to invest twenty-six billion dollars in Airbus Industries to dominate commercial aviation and displace the U.S. The strategy is working. (Note that we said invest rather than subsidize. It appears that the Europeans will do well on the money as an investment, plus getting the taxes from all those profits and jobs.)

Management Models: From Control to Empowerment

We have talked much about emerging differences in effective management style, but have avoided predicting exactly what the future successful company will look like. The reason is that there is no one best model for success. In the information age, as with architecture, one best designs by keeping in mind the materials to be used and the location and application of the building. "Best" is relative, and less can be more.

The *keiretsu* form of business works well in Japan, but probably will not transplant outside that culture.[12] The *chaebol* form of business works in South Korea, but it certainly will not transplant to the West.[13] The law firm style of organization works for Microsoft, but it probably will not outlive the energetic life of Bill Gates as a personal contributor. Few industry CEOs possess Mr. Gates keen understanding of technology, user needs, business, law, and poker playing. Almost none possess the ability to transmit a clear vision of a future that allows in-depth motivation for a multibillion dollar company.

It is clear that the old, industrial age, silo organizations so common today will decline. But, just as millions of years later many species of dinosaurs survive, so will it be with the old top-down control functional forms of organization. Within such hierarchies of managers the key talent is that of career advancement, and this inhibits competitive advantage. As long as democratic governments persist into structuring as functional bureaucracies, it will drive industry to form similar structures. But so what? These relics of the past will not be the vital organizations or organisms of the late '90s. They are not worth discussion.

The industrial ecology of the next century will consist of a myriad of corporate life forms, some of which we can't even imagine today. When the dinosaurs faded, they were replaced with a rich ecology. The dominant life forms were mammals, and most of these were small, little noticed, but very successful predators like foxes and raccoons. Thus it is in the information age. Micro-capitalism is flourishing. Individuals are starting unique knowledge-based businesses, many of which will look dramatically different from any businesses of the past.

From accounting, computer service, and finance to musical instruments, software and real estate, such micro-businesses are the fastest growing segment of the U.S. economy. These are enabled by technology and encouraged by the slash and burn downsizing trends of financially based large corporations.

These new micro-businesses are interesting, but not worth discussing in this book. No two will be alike. Each is formed around a core group of skilled, experienced, people, and each carefully targets activities to piece together enough income to survive at a desired level. These micro-firms are all privately held and often unwilling or unable to disclose their practices in detail. Even if such firms reveal how they function, the details are so highly specific to the capabilities and interests of its core group as to not be of general business interest.

The Trudel Group (TTG) is an example. It is a management consulting firm, but one ever so different from the mega-consultants like McKinsey or Boston Consulting Group. TTG is a boutique serving selected clients with custom work, mostly in the areas of strategy and product and business creation. We have only one employee. Even our accountant, legal, and secretarial staff are independent subcontractors, but we can tap a network of world class knowledge experts. We call what we do "business innovation."

TTG is, in effect, a "virtual corporation." No two assignments are alike. Each is designed around a unique statement of work to deliver maximum client value. In fact, we are bound by the Institute of Management Consultants' code of ethics to always serve our client's best interests. The large management consulting firms, in contrast, must emphasize mass production to keep junior workers backlogged with billable hours, so they tend to offer standard process and products that can generate large revenue streams. Some bundle their services with the reselling of products, thus perhaps making it harder for them to give objective advice.

The more interesting question is: What will the large, successful, Western corporate predators resemble? In essence, what life forms will become the lions and tigers of the information age ecology? In specific, what will be the successful management practices of such firms?

The answers to these questions are slowly becoming clear. Organizations like Hewlett Packard, Intel, and perhaps even GE and Microsoft give clues. The similarity is not so much organizational structure as it is that these are all flexible, fast-learning organizations.

There will be rules to guide the conduct of large corporations, but these will be very simple, perhaps one page of paper. HP fit their new guidelines for new product development into a list of six (6) items. Intel has a shared vision guided by a roadmap, a single poster that sets corporate direction. Moore's law, a single graph, dictates how fast each business must accelerate the performance of its products to stay in a leadership position. GE's CEO, Jack Welch, has noted, "For a large organization to work, it must be simple."[14]

Within these guidelines, which communicate strategic direction and core competencies, local managers and professionals will have considerable latitude to act. Business entities will be small, a few hundred people or less, nimble, and free to leverage the full power of the enterprise. These will be further divided into teams, built around three things — leaders, a mission, and attachments between members.

The managers and leaders will be responsible for adding value, and will be judged by decision quality. Teams will be judged as teams, and based on the results achieved. However, the definition of "results" will be broader than today's financially-based scorecards. The key sustainable competitive advantage will be the ability to learn faster than competitors, so a team that secures vital knowledge for the corporation may be deemed successful even if traditional business results, new product profits for example, are lacking.

Teams will often be extended, including trusted members from outside the business entity or the corporation. The emphasis will be to compress time and exploit the best knowledge. That implies use of inside and outside consultants, and also strategic partnerships. These trends are already well established.

Rather than disciplinary controllers, team leaders will be along the old Roman model, *primus inter pares*, first among equals. The function of team leader may rotate and may be shared. It may be influenced, but probably not dictated by, the team members. Teams will train together, and develop mutual respect, fierce loyalties, and commitment to results. This can, of course, only work in an environment of mutual trust.

Honest mistakes will be allowed, and, in fact, learning through inexpensive failures will be encouraged. Most likely the punishment for mistakes will be the recognition that one let the team down, plus self-obligation to atone or set matters right. Because corrupted information is lethal, and because verification delays are unacceptable, dishonesty and deceit will cause rejection from a team culture. The teams will be self-managing and empowered to act as they deem necessary within the corporate mission.

Work will be integrated, not fragmented. Power will be expressed both via managerial authority and through knowledge and experience. Authority will be used to remove internal blockages, but knowledge and experience will be used to set direction. Some think of this as analogous to hospitals, where the skill of doctors is enabled by an administrative structure that provides funding and facilities.

There will be organization and structure, but it will change with time and will be arranged more along key processes that add customer value than around functional specialties. Jobs will be time limited and results based. There will be a duration of assignment, and over this period things will get done or get better. Assignments will be focused so the results can be measured and it is apparent when the assignment is over.

If successful, the person will then move to another assignment that will enhance her value to the corporation. If unsuccessful, she may be moved to a role that lets her learn and offers another chance for success. Over their careers, the future leaders of the firm will be expected to experience both success and failure. People will be loyal to the company because it is loyal to them, not because they have "job security" or are obedient to a senior bureaucrat.

From Traditional Silos to Cross Functional Teams

You can easily identify traditional organizations. Structures are vertical, not horizontal. Organization charts resemble silos. There is preoccupation with problems, blame, and near term events. Managers worry about the new budget cuts, the rumored reorganization, the recent competitive product announcement, or the slips in schedule for the new product. Discussions of shared vision, autonomous teams, and empowerment are rare, and usually lip service for slogans or annual plans. Cheap labor is preferred to workforce skill development.

Capricious management edicts can create severe legal liabilities for an organization, so "make the boss happy" has been largely replaced by "adhere to company standards and procedures." It is now safer to fire people because they violated a standard procedure about keeping workspace tidy, or whatever, than because the boss does not like them. To deal with "problem" employees, simply enforce all the rules all the time and they will eventually acquiesce or provide grounds for termination.

Traditional organizations cherish a minutia of metrics. It often seems that it doesn't matter what is being measured, if management is measuring it, on top of it, and fixing the problems. Workers are overworked robots, and managers scamper around with their hair on fire, computing variances and solving problems.

Significant Datum #1: If one works for a Japanese company, colleagues will interrupt you when they see you engaged in action but leave you alone if they see you quietly thinking. In U.S. firms the opposite is true. People interrupt when you are thinking and leave you alone when you are pounding problems.

The reason for this is that the two cultures have disparate perceptions of where added value lies. In one case, thought and knowledge (content) is seen as valuable, so workers treat quiet time as sacred and avoid interruptions. In the typical U.S. business, it is action (process) that is deemed important and not to be interrupted.

Significant Datum #2: There is a high correlation between product success and front-end investments for knowledge. Front-end work is planning and gaining knowledge. This design work is absolutely crucial for product success, but is usually a dim and fading memory by the time the product is launched. Frantic problem solving activities come later, and these usually preoccupy management.

Average U.S. companies spend about 4–6% of their product development funds at the fuzzy concept stage. Leading firms spend 15–20%, and some Japanese companies spend as much as 25–30%.[15] The implications of these disparate investment patterns merit serious thought.

Japan is famous for sensitivity and planning. They advocate two distinct modes of time. They take as much time as is needed to gain deep understanding, but then implement quickly. The U.S. is known for simplistic but direct action. The hero rides in with blazing six guns, shoots the bad guys, and problems are solved.

Western managers have a proclivity to charge off randomly, intending to prevail by fixing any problems as they arise. As one client told us, "We think with our swords and shields in our hands." These people often jump from method to method without ever pausing to understand the situation or why previous solutions failed. Business reorganization is continual, but objective post mortums are strikingly rare.

The best examples of narrow, mindless, problem-solving processes are found in government organizations. One friend was a laid-off airline pilot. She applied for unemployment, and grew concerned when many months passed without any payment. She called to ask what was causing delay, and got a manager who explained they were understaffed and, thus, far behind on their claims processing. Desperate, she pleaded for funds to buy groceries.

The manager was sympathetic. He prodded the system and action resulted. A clerk called, but our friend was out looking for a job so all that resulted was a message on a recorder. The next day my friend called back, but the phones were busy. She finally got through, only to learn that her claim was rejected. She asked why, given that no one had even talked to her. The clerk tersely explained that "the system" only required that she be called twice, and that the clerk had done so and gotten a busy signal on her second and last call. Hence, the clerk had done all that was required. Procedures had been followed.

Consider the economy and efficiency of such a system. With simple written procedures, one does not need skilled workers. That clerk is processing claims quickly. Neither she nor her boss is happy, but they are good flesh robots. Either or both can be easily replaced, probably with cheaper help. New employees might churn the system faster, at least for a while.

Actually, of course, the effectiveness of such an organization is very low, as with most bureaucracies. Friction is maximized and output is minimized while generating impressive data to "prove" good performance. The game is to blamelessly follow procedure and push problems over the horizon to another part of the organization.

The people on the other side of the horizon, of course, do the same. Thus everyone "fixes" their problems while workloads and staffing levels grow without limit. The typical organizational structure in almost all firms is vertical. It resembles a collection of functional silos (some call these chimneys). Functional activities do not come together until one reaches the general management level.

Dissection and separation work poorly for information age problems, as has been pointed out in such classic works as *The Mythical Man Month.*[16] It is well known in the software industry that adding workers to knowledge problems slows progress and that making decisions without close, timely, intimate knowledge is dysfunctional.

In today's key markets product development and starting new businesses are also uncertain, time sensitive, complex, knowledge -based systems. Viewing these as activities and separating them into pieces doesn't help. Instead it makes it impossible to competitively reconstruct the whole.

Making executive decisions in today's business environment is not much different from making decisions for subtle, uncertain, and complex software projects. Perhaps that is why a hard driven hacker like Bill Gates can do so much better than the fully supported, process trained, resource rich managers at traditional firms.

Box 7-2

Interesting Data: In recent years far fewer CEOs have lost their jobs or resigned from software firms, though stress in that industry has been as high as in computer hardware.

When you break an elephant into small chunks, you do not get a set of small elephants, you get a mishmash of dissimilar parts. Instead of little "Jumbos" you get big jumbles. It is simply not possible to understand complex issues by looking at single, disconnected parts. But understanding the whole as a system, and quickly, is the key to winning in knowledge-based industries.

Consider if you took the best parts from the best cars in the world and put them into a warehouse. You took the best engine from one car, the best suspension from another, the best transmission from a third, the best dashboard from a fourth, and so on until you had all the parts needed to build a car.

When you went to build your car, you would find that you could not construct a functional vehicle. You did not have a car in your warehouse, you only had a collection of good parts. That is the *second* best result you can hope for from disconnected activities. The best result, a kludge of components that partially work together, won't win against first rate competition either.

Then there is the issue of time. For products to reach the market they must successfully flow up and down all the parallel silos in a traditional firm. In today's key markets, the first competitor with the "right" product usually wins. It is easy to show that time to market (for the right product) has more impact on profits than other factors.

Less discussed than time to market — but perhaps more important — is "time to money." A firm launching a new product or business is vulnerable until they reach break-even. Getting to market fast is a good practice, but getting to profitability rapidly is even better. That means the production and demand generation must "ramp" very rapidly.

Silos fit the "skinny arrow" project model. See Figure 7-2. Work is sequential, and each group tosses results over the wall to the next. The group on the receiving end does their best to cope, and if they cannot they toss it back for rework. In the 1950s and 1960s, decade-long product cycles were common.[17]

The "fat arrow" model developed by Trudel in the 80s, see Figure 7-2, depicts the concurrent, empowered, multi-disciplined teams

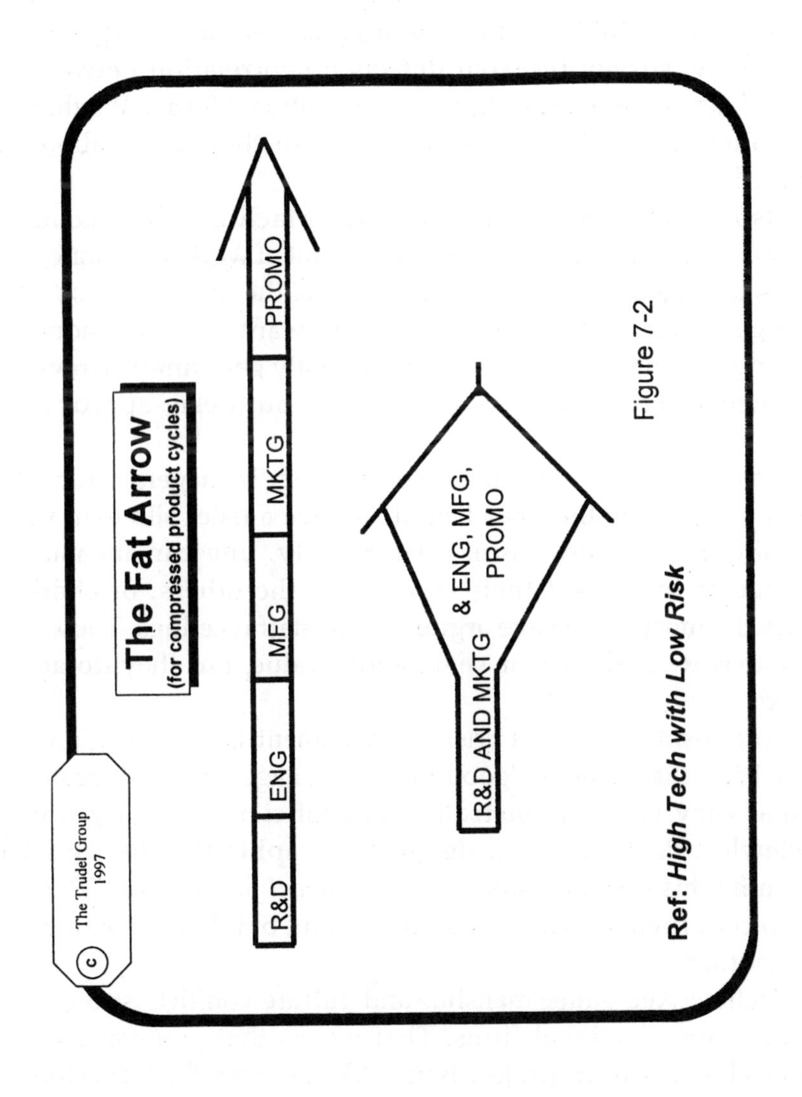
The Fat Arrow
(for compressed product cycles)
R&D
ENG
MFG
MKTG
PROMO
R&D AND MKTG
& ENG, MFG,
PROMO
Ref: *High Tech with Low Risk*
© The Trudel Group
1997

Figure 7-2

that are so universally espoused today.[18] Most firms say that they now use this form, not silos, but the results don't bear this out. It appears that only about 10% of work is truly concurrent. That was baffling, given that the fat arrow approach is clearly superior. Still, even today, survey research detects no correlation between new product development results and organizational forms. We think that is because companies are not honest with themselves about such things.[19]

Also, concurrency conflicts with Machine Age accounting systems. The area of the arrow is the amount of work to be done, which relates to project funding and staffing. As the arrow gets shorter, it must get fatter. Still, accountants often want to "save money" by delaying staffing. Hence, firms inadvertently get skinny arrows. They flog their project managers relentlessly, but delay approvals and funding.

Such behavior should not be surprising. Managers can calibrate career progress by their headcount: "I have a taller silo than you do, so I should be paid more." Alternately, one can measure the prestige or budget of their silo versus the others, or chart who makes the most persuasive arguments in staff meetings. These games do not create customer or shareholder value, but they do advance careers.

More insidious is that crisis management is so highly rewarded in the West that, even in "good firms," managers tend to create crises so that they can demonstrate this proficiency. Randy Englund of Hewlett Packard has a delightful graphic that he uses when presenting his book, *Creating an Environment for Successful Projects* — it depicts a fireman with a hose in his hand but flames coming out of his mouth.[20]

Machine Age gamesmanship and culture conflicts strongly with the new organizational firms. That means that you can't separate top level issues from project issues like new product development. To say it once again: A holistic view is needed. Machine Age firms are quite robust, and transformation, not incremental change, is required.

Weak Signals Presage Opportunities

John Akers of IBM preserved profits by investing in mainframes, but missed emerging microcomputer opportunities. John Scully did well selling Macs, but missed major opportunity in the late 80s by failing to port Apple's seamless graphical operating system to Intel based PCs and calling it Windows. Scully, outmaneuvered by Bill Gates, not only licensed his OS to Microsoft but canceled an exceptional product, MacBASIC, signaling that Apple conceded software and was content to be a platform vendor.

Today the basis of competitive advantage is timely knowledge, and this makes the price of risk aversion extremely high.[21] It just is not possible to acquire deep knowledge quickly without experimentation and failure. Consider that the most successful technology based products usually come after several cycles of "failures."

Apple's hugely successful Mac grew from Lisa, a failure but an innovative product. Intel's 386 and 486 emanated from the substandard, brain-dead 286. Lotus is betting the future of its company on Notes, a product which has been losing money for almost a decade. Motorola's most successful businesses were once misfits that were saved only by leadership and intervention. It is an industry joke that Microsoft has to do products three times before they even work. The average *successful* high tech new venture is cash negative for seven years.[22]

Incidentally, one other lesson from the Mac is the high mortality rate of product champions in Western firms. The early Mac visionary and leader, Jeff Raskin, was fired for complaining to Apple's Chairman that Steve Jobs was a "dreadful manager." Jobs, who reportedly initially opposed the Mac, later became the evangelist who launched it with religious fervor. Jobs in turn was fired by his friend, the mass marketer Scully, who later "resigned" himself as other talents became more needed.

A Japanese firm would recognize that all these roles and contributions were necessary, and they would take steps to ensure that their firm retained such talent. Though misunderstood, their policy of lifetime employment for core employees and takeover protection for firms offers major competitive advantage.

As we move toward "progressive" management, we move further from leadership and innovation. Consider the normal practice of rewarding managers for "results." In theory this sounds reasonable, but leaders and organizational politicians know in their hearts that it has flaws.

Some wily rising executives groom sycophants to consolidate power. Take your best sales territory and give it to a henchman. Give the non-ally who developed the territory a new challenge, perhaps Siberia or Montana. You not only save on sales commissions, you develop bullet-proof reasons for promoting your protégés by assuring that they are associated with success. If the non-ally succeeds you got hard work done cheap and can give him another challenge. If he fails, he may leave and you can pick his replacement.

The military calls this grooming process "getting your ticket punched." The practice is common, and IBM reportedly did it in depth before Gerstner's first downsizing. Managers in power gave themselves and their allies useful jobs and put outsiders or enemies in redundant jobs before the layoffs.

Another version of this story is the general manager who inherits a superb division. The products are compelling, the market is growing, and a well-honed team is in place. If he does nothing at all, performance will be good for years. If he guts it out, cuts investments, and lays off expensive talent he might make it look very good indeed.

This too was common in the '80s. Manipulators, accountants, and lawyers — not builders — seized power in most U.S. firms. Charisma, personality, and numbers were more important than innovation, market share, and the character ethic.

Only a very few leading firms recognize the problems with results-based management evaluation. Hewlett Packard has started evaluating management performance based on the quality of decisions made rather than their business results. This takes a talented and honest organization, for only dispassionate review by competent peers can assess decision quality. Intel also coaches to raise decision quality. Both firms forgive learning errors, but neither tolerates stupidity or chronic blunders. You know you are working with a first rate firm when a senior manager or professional dares say, "Yes, I made a bad call. It cost us four months and $500,000."

From Flush Solutions to Mindful Action

In industry, the "fix problems" bureaucracy often sets targets against industry averages. "If our competitors spend 10% on R&D, then that is what we should spend." You can bet that at lower levels someone somewhere is measuring things like the key strokes made by secretaries or the number of sales calls made per day. Presumably, they will then compare these to standards, and fix deviations. Is that the road to competitive advantage?

There are widely reported frustrations over measuring the productivity results of workgroup computing because experts disagree about metrics. How can all that spending for computers and communications be justified? They know counting meetings or memos is silly, but feel a compelling need to measure *something*. Much money is being spent to compute return on investment (ROI).

The answer is obvious, but it doesn't fit into simple cause and effect problem solving. Every competent person on the planet knows that firms that fail to exploit modern information technology are noncompetitive, but in the U.S. metrics and financial calculations are still needed to justify action. Therefore Singapore is 100% wired for ISDN, while the U.S. is only 2% wired (fibered, actually). Officials talk of fiber highways, but most of the country lacks paved roads.

To fix problems, the traditional manager's goal is to remove all possible personal, personnel, and subjective factors from the system. In such a Nirvana, the workers are cheap and interchangeable. One can define process, measure the results, and fix problems ad-infinitum. This is classic Taylor process, vintage circa 1900.

Managers want predictability and order, and drive toward it relentlessly. They think people are the problem, not the system.

While there is nothing wrong with process, there is a lot wrong with bureaucracy and systems that prevent people from using their knowledge and skill to add value, change, and do their jobs better.

Consider the "fix it" techniques that have absorbed so much attention for decades: strategic planning, competitor intelligence, market positioning, loving customers, benchmarking, BCG industry matrix ("Cows, Dogs and Stars"), centralization, vertical integration, decentralization, Strategic Business Units, PERT, econometrics, experience curves, critical success factor analysis, Kepner Tragoe, marketing warfare, flexible manufacturing, concurrent engineering, CIM, the one minute manager, management by objectives, inventory turns, creative financing, reverse engineering, PIMS, MRP, MRP II, zero defects, zero based budgeting, the management grid, theory X, theory Y, theory Z, market gap analysis, employee involvement, and the list goes on endlessly.[23]

More recently, we've had downsizing, rightsizing, outsourcing, re-engineering, strategic partnering, time based competition, Computer Aided Everything, and the myriad of Japanese based quality techniques. We had quality circles, zero defects, JIT, committed to excellence, TQC, TQE, TQM, QFD, six sigma, Tagauchi methodology, Continuous Process Improvement, core competencies, and many more.

It is sometimes amusing to watch how these methods come and go. Ten years ago, a U.S. firm we know had an engineering culture and archaic manufacturing processes. If a product was defective, it took marketing a year of intensive effort to get a "mod" processed. It took months more for manufacturing to use up its old parts before the first product emerged with the problem corrected.

Then came the era of "loving customers," and a Vice President took up the challenge of fixing this system. He succeeded, and soon defects could be corrected in days or weeks. Customers were happy, and the fortunes of the firm improved.

The Vice President was still there when fast cycle time became fashionable. This area was no longer a problem, but this was a process that could be speeded, so speeded it was. If a customer even flinched, mods were processed instantly. No one knew or cared if extra speed added value, because that wasn't the point.

Next came the era of cost cutting, the Vice President was replaced, and most of the experienced mod processors, being senior and well paid, were laid off. The system started to falter, and the next era, quality process, brought it to its knees. Now all processes comply with ISO 9000 and are detailed and documented. Presumably millions of mods can be processed before the first error occurs. Does anyone really think that level of perfection adds value?

The main result is that after a decade of work, the time to process a mod is snail-like and almost back to what it was at the beginning. Since the mod process is now imbedded in a company wide quality effort, and because managers are terrified and risk-avoidant after years of downsizing, it is unlikely to improve soon.

Dr. Joseph Juran, the quality maven, commented in his retirement speech on the incredible proliferation of methods. "If I want to be mean, I'd say we have situations where some people want to be on top of a hill. Alas, every hill has a king. What is more logical than to create a hill?"

For the past two decades, we have shored up classic management with several new processes and techniques per year. We have moved to fixing things at a higher level of detail, faster, and to ever increasing standards of quality. Most firms moved further from deep understanding or entrepreneurship in the process.

So far not one of the techniques named — not a single one — has delivered lasting competitive advantage. Though all had merit and helped in a few cases, most only improved matters for a minority of the firms that embraced it and were soon abandoned. Like waves on the beach, a new process appears, sweeps the landscape, and soon vanishes with little to mark that it had ever been there.

From Process Approaches to Action Experiments

In the U.S. we tend to see "process" and "management" as interchangeable words. We see learning as "exposure to information" (e.g., reading a book) and we don't place much value on content knowledge, experimentation, or practice. How often has the U.S. government done trials or Beta tests before rolling out massive, expensive, bureaucratic programs? Probably never.

Western corporations are almost as bad. For decades and until near death, most resisted actively tapping "outsider" knowledge like consultants, customer involvement for product improvement, and worker involvement to improve quality. Even today, our companies have the equivalent of sports team owners, managers, and trainers — but they lack coaches. We resist practice and often don't bother to keep teams together. We know that neither war nor 0sports events ever play out as planned, but seem to think business competition should. Does anyone recall what General Schwarzkopf did when told to win the gulf war? He assembled his troops, put logistics in place, and had them practice as teams for six months. Then he won the war.

Stranger yet, Western managers seem to think that picking a method from the alphabet soup of processes will somehow cure all their problems and compensate for a lack of training and knowledge. That makes about as much sense as a golfer who has one favorite club that he uses to the exclusion of all else. As with golf, a selection of good clubs helps, but the skill of the player makes the most difference.

A quest for "standardization" infects the consulting professional as well. One of the authors was rejected by a certification board for confessing that he provided custom work and valuable results, not standard process, for his consulting clients. This resulted in supportive reaction by colleagues, heated arguments, and an eventual reversal. Many traditional managers sincerely fear that any movement away from standard documented process is the first step on the slippery slide to anarchy.

Ironically, during this debate the *Wall Street Journal* ran a front page article highly critical of a large, prestigious, international management consulting firm.[24] Critics said in part, "Their work is canned, their formats are canned, and they give black box solutions to everything, where every situation is unique."

The firm in question bills well over a billion dollars per year, and it is typical for such firms to charge over $1 million per assignment. Recently, many accounting firms have moved into consulting, seeing the market as a growth opportunity. Some of these serve, in effect, as large value added resellers (VARs), often not only providing counsel but complete turn-key computerized database systems to automatically run your firm. (Conversely, members of The Institute of Management Consultants in New York, are bound by a written code of ethics to avoid conflicts of interest like selling equipment or services in addition to advice.)

Still, criticized or not, standard process is popular and lucrative. Clients buy it. The "one size fits all" mass market business model is typical, and successful, for most large consulting firms of the Machine Age, if not always for their clients.[25] The mega-gurus draw similar criticism and financial success.[26]

At least one large ($250 million) consulting firm was widely reported to have manipulated the best seller list so as to include their book.[27] Consultants fortunate enough to become associated with best sellers (Yes, please do tell your friends and colleagues to buy this book!) are richly rewarded. Those with popular media "buzzwords" can get five or six digit fees for lectures, so the big names and large firms are now suing each other over slogans.[28]

The conundrum of generic solutions versus custom products comes up at all levels and in all markets. In past eras mass production won, but today the economic build quantity is one. Customers or clients can get products that match their exact needs. From consulting to computers, the leading brand is "other." Even in markets where some particular brand name dominates, close examination usually reveals an ability for the customer to get exactly what he needs.

Unfortunately, Western thinking is geared to generic solutions. A colleague is a marketing manager in a high tech firm. Recently he did a survey of customers, and found highly favorable customer perceptions that he wanted to publicize for his employer's benefit. He was overruled because the firm had not yet achieved "total quality standards" for his marketing efforts.

We must be missing something here. Happy customers without applicable quality standards are deemed to be of little value? Wasn't making customers happy the stated purpose of quality?

Many intelligent people — most of them, in fact — have embraced using some particular process for generically fixing broken things as a core capability. To oppose fixing things seems like opposing Quality, and that seems absurd, doesn't it?

The truth is that all this problem fixing through generic process — actually does improve things. But only for a while, in stable situations, against the right competitors, and if the right process is used appropriately, in context, and not to excess. And only if the firm keeps a core capability of innovation to feed process. (Perhaps, as with medication, we should affix warning labels to management technique?)

It is becoming increasingly common for entire industries to drift into mediocrity and then be destroyed by competitors who change the rules and paradigms. The problem with a core strategy of fixing broken things is analogous to driving down the road by watching only your rearview mirror. If you practice diligently, you might get very good at the technique, but an adversary who looks forward has major advantage.

Attractive industries attract the best competitors, and you can bet that some have progressed beyond simplistic problem solving. If your guidance system is backward looking, inevitably you will someday miss major new opportunity or competition will place a boulder in your path. Even the best process all too often can lock one's gaze to the mirror of the past.

When problem solving technique fails, as it does inevitably, we abandon whole industries. Since the fifties, the spectacle of Western firms exiting lucrative, growing industries has become common. It started in motorcycles and steel, then automobiles, and finally high technology.

"You can't make money in consumer electronics, better move to semiconductors." "You can't make money in semiconductors, move to computers." "You can't make money in computers, move to software." "You can't make money building anything, so give up. Source from Japan (or wherever) and become a sales and service company."

The U.S. has, by political choice and trade policy preference, suffered high wage job loss in the manufacturing sector, and massive, ever-growing trade deficits, especially in technology sectors. Bills now in Congress to weaken the U.S. patent system and other intellectual property protection will, if passed, pretty much ensure that most of the rest, except for some assembly operations, is moved offshore. Conventional Machine Age economic theory says that this does not matter.

Still, the authors respectfully suggest that these policy decisions deserve extensive debate. We think that many experts now agree that future economic and corporate prosperity will come from building products with high value added knowledge components.

Box 7-3

In the long-term, the U.S. will become the first 'deindustrialized' country, defined as having less than 10% of economic activity in manufacturing.

Dr. Percy Barnevik, Chairman and CEO
ABB Asea Brown and Boveri, Ltd.
International Industrial Conference, September 29, 1997

Innovation Versus Refinement

The biggest problem with a strategy of fixing problems in today's world, is that the best markets are turbulent and competition is so resourceful and competent that it is sometimes safer to drive change than to react. Recall the analogy to driving by watching the rearview mirror, and ask yourself, "Is it possible to ever become a first rate company by simply fixing problems and improving current processes?" If all you do is fix problems, how can you ever move to the next new technology? How can you create new markets? How can you develop vision for what might be if you are consumed with short term reaction to other's actions?

Certainly, if your firm is performing at a low level, you can make dramatic improvement by simply copying others. If your order processing is a disaster, benchmark against a firm that does things well and copy their methods. If quality is a disaster, benchmark against a firm that does well and copy them. This approach works for awhile, but soon reaches the point of diminishing returns.

The next step is to not only benchmark, but to copy other's improvement processes. Copy quality circles, copy TQM, and perhaps then you can improve too. That, for the most part, is what the industry has been doing. This also quickly reaches the point of diminishing returns.

Let's look at the results. First, some imitators do so poorly, and these fail. We will discard these. Some are good learners and achieve good reverse engineering, TQM, cycle time improvement, or whatever. What is the result?

There is a growing body of evidence that the result of all this copycat activity is short term gain followed by drastic loss of industry profitability. When everyone in the industry has good quality, it is no longer a competitive advantage, just entry stakes to play the game. In these cases, prices and margins erode as the industry desperately tries to keep plants busy and the workforce employed.

In mid-1993, which firm cited in the classic *In Search of Excellence* had one of the best balance sheets in the computer industry? Answer: Wang, the firm that — imploding, disorganized and just emerging from Chapter 11 — missed *all* the trends, but evaded liabilities. They failed at new products and downsized from $3 billion to $1 billion, slashing manufacturing employment from 9,000 to only 100 workers. This is winning?

The personal computer industry has many talented firms, and most have perfected benchmarking and copying each other's processes. They all solve problems well. Most in the industry are losing money as they thrash through faster and faster cycles with the same undifferentiated commodity products. Not only are margins razor-thin, but one slip or mistake and you can lose a whole product generation and never recover.

Watching clever firms fail on such a broad scale leads industry observers to damaging conclusions. It is now common wisdom that, since you cannot gain advantage from new products, one should concentrate on better processes. This is half right. In reality, for a firm to benefit from a new product one must be early to market with high quality and low cost. That takes both innovation and process.

It reminds one of the Western exodus from consumer electronics two decades earlier. Apple plans to let the Japanese build Newton, their next product generation. Dell is considered a PC industry leader, despite being, in essence, only a distribution and service company. Many other Western firms are moving to service strategies for mass market PCs and interactive TV devices. We suggest that such migrations are the artifacts of corporations being driven from their homelands by more formidable competitors.

Every first rate firm the authors have studied does far more than problem solving and process improvement. The leaders all have a broad vision of what the future can be. Intel exploits the ever increasing power and density of silicon. Microsoft exploits the pervasiveness of software on the desktop. Motorola exploits the information age's need for communications. Sony exploits the impact of electronics on personal lives. All of these firms use vision, skill, knowledge, and, yes, process, to ensure that as the world they envision develops into reality they have the core competencies to allow competitive advantage.

Top managers are starting to complain that "all this 'team stuff' is expensive and it's not working." The complaint of a V.P. of manufacturing is typical, "We used to have to train a single supervisor, now we are required to train everyone." Such is life. New technology — teams should be thought of as a new business technology — usually requires investment and incurs a period of lowered productivity.

That's a problem. Managers like control, and teams are a threat. Even Peter Drucker has little hope for a happy and tranquil evolution into the future. "Gradual change cannot work. There has to be a total break with the past, however traumatic that may be."

There is growing consensus that before teams can work, their parent organizations must change. The key requirements include (1) clear communication of strategic goals and priorities to teams, (2) changing functional managers from controllers to suppliers, (3) making teams, versus individuals, accountable, (4) investing resources, and (5) giving professionals authority to choose assignments.

Doing any one of these things is very difficult for most firms, and doing all takes years of strong leadership. Most Western firms won't adapt to collaborative capitalism in time, but a few, perhaps 10%, will be successful. These will tap outside talent, and their leaders will recognize that past experience is no guide for the future.

The successful managers in the leading firms of the 90s will behave very differently from traditional managers. The role of controlling people and staying on top of all events will change to a focus on guiding, energizing, and exciting those who work in the organizations they manage.

Even today, in the leading edge firms, the traditional questions are meaningless. If you ask an employee of such a firm how many subordinates report to his manager, you are likely to get a frown. "Sorry, you seem to have it backwards. My manager works for the twelve of us to help us succeed at our jobs."

The legendary University of Alabama football coach Bear Bryant captured the essence of this leadership style in a saying about recognition. "If anything goes bad, I did it. If anything goes semigood, then we did it. If anything goes real good, then you did it." Middle managers will increasingly become load bearing structures. Their focus will be mostly internal, and their mission will be direction setting and enablement of the teams that report to them.

There will be few clear-cut orders from the top. Rather, information age managers will have to have a very clear understanding of corporate strategy, its implications, and the capabilities needed to achieve it. Rather than following policy and procedure, their role will be improvising as the immediate situation requires. They will focus on taking care of their subordinates, and will elicit from them knowledge of what needs to be done.

This implies key roles for content consultants and professionals. They will be the knowledge experts, their focus will be external, and they will learn continually and at an increasing rate. Marketing professionals will know their customer and competitive environments intimately — not so much with numbers, as with more subtle types of understanding. The same will hold for technical professionals in other fields, such as engineering, science, and finance. There will be substance in the "dual career ladder" that is only given lip service today. Professionals will increasingly be expected to grasp external markets and capabilities *before* they become obvious.

The rewards and emphasis will be on results, not on methods, process, and image. This means that jobs will have less structure in space, but more in time. Much as an outside consultant will contract today for deliverables, so will managers and professionals in the leading firms of tomorrow. Rather than managing the "microprocessor division," one will lead the "XYZ development." There will be time lines where things are expected to get done, get better, or both.

Rather than having a "job description," one will have a portfolio of projects. Teams and portfolios will be dynamic, and nothing will be taken for granted. Projects will end or be canceled, perhaps often. For this to work, of course, failure must be allowed, but not stupidity or deceit.

Winning firms will rapidly apply the best talent available, and that means both internal and external sourcing. Rather than rewarding managers for headcount or budget, or giving individual performance reviews, rewards will be team rewards based on results. If one does wonderful technology in a failing product, the reward might be less than for adequate technology in a winning product. If one leads a small high risk new venture well, one might get more reward than for presiding over a mature, large, profitable division.

Organizations will be "virtual" but the elements must have close, interdependent, high trust relationships. The team might need a market plan, not a market planner. They might need a product definition, not a product definer. A brochure, not a brochure writer. In such cases they will engage "outside," but trusted, help.

Focus will shift away from organizational size and stability to choosing what works best for the situation at hand. Knowledge and skill gained by continual practice will take precedence over corporate power, process, and edicts.

Customer Surveys to Enlightened Discussions

Most of what is taught in marketing comes from the industries that invest the most in marketing and promotion, consumer products. Compared to high tech, much more of what makes or breaks a consumer product is raw positioning and promotion. It is worth major promotional investments to shift a few points of market share for a soft drink.

In the authors' experience, the marketing tools developed for traditional products have never worked very well for the emerging, dynamic markets that are so crucial today. Some of the biggest failures we've seen were associated with elegant market segmentations and reams of data to "prove" how good the product fit is to customer needs. Conversely, some of the best products we know of — e.g., Sony's Walkman, Apple's Mac, Intel's 386 — could never have been justified by conjoint analysis or quantitative, statistical market analysis.

Why this is so is debatable, and some data exists that the classic tools are losing their power even for the traditional markets. For example Pepsi's advertising campaign won awards and tested superbly well at audience brand awareness. By normal standards, their promotion was excellent, and by contrast Coke's promotion and positioning was woefully inadequate.

But Coke gained market share and Pepsi lost. How can that be? Because Coke marketed to the food chains based on relationships, and one Burger King chain can consume quite a bit of Coke.

Anyway, high technology marketing has always been intuitive, and most successful professional practitioners know this. Technology-based products are very subtle things, and the most useful knowledge is gleaned from high content discussions with the "right" people. Talking with a few visionary experts can be better than collecting statistics.

If a person buys a personal computer, they typically spend much time reading reviews, talking to experts, and researching technology trends. Some of the best market segments are power users who want the "hottest boxes," but will fax twenty sources and take the lowest quotation. They almost never visit a computer store, because they know more than anyone they are likely to encounter there. They tend to ignore or discount advertising claims.

No one would go to that much trouble when buying a soft drink, and few do it when buying cars, but such a selection process is common in high tech. If the purchase involved is a LAN for a company, or a fleet of aircraft, or a super computer, the selection criteria become even more rigorous.

What matters in such markets is content and knowledge, not statistical data. One smart, informed person who understands what way technology and competitive forces will lead the market outweighs a thousand uninformed or misguided survey points.

There is a saying in the computer industry, "The only way to make a lot of money is understanding a customer need before the customers are aware of it, and then bending technology to meet that need in a way that is uniquely yours." The saying is true.

Casey Powell, the founder and CEO of Sequent Computers, also has football analogy we like. "To play in the major leagues you have to throw the ball before you see the receiver." This has several meanings, all of them helpful. For one thing, the quarterback cannot stand back and "manage" the team in typical industrial age business fashion. If he does, the team will lose. Instead he has to do his job well and he also has to trust his well-trained team.

Recall, we in the West mistake "trained" as synonymous with "exposed to knowledge." For example, "I took a class, read a book, and got a certificate. Now I am trained in digital circuit design." The authors' definition of training goes beyond this and requires the actual repetitive practice of knowledge. One becomes a concert pianist or a professional sports player only through a combination of learning and constant practice against increasingly more difficult competitors.

The quarterback has a difficult task. Adapting in real time and after misleading the defense, he throws the ball, hard and accurately. Then, if everyone on the team makes their blocks and runs their patterns perfectly, there will be a receiver in the right place to catch the ball and he will have a three-step lead on his opponent. Success depends on skill, team behavior, trust, and risk taking. So it is in business in the dynamo markets.

In the computer industry, one must usually start product development before anyone *knows* and can prove the market need. More important, however, is that every one of your opponents is smart and well trained. They all read the same surveys, they all attend the same conferences, they all have good technology. The winners move beyond quantitative data to qualitative data, and the process is enlightened discussion with the right people. Eventually everyone will target new business opportunity, but the winners are those who can consistently do it early and well.

This is art, not repeatable bench science, and it can be much fun. There are no rules, and the technology and markets are inexorably intertwined. The process of creating a winning product is subtle, high content, interactions between team members, and extending to customers and suppliers. Winning depends on cross-functional knowledge and enlightened discussions. Successful practitioners generally combine both technical and business school training, and years of apprenticeship and experience.

Incidentally, least you think we suggest that only the big firms have problems, that is not at all true. No matter who you are, you can suffer attack from both larger and smaller firms. The software market is unusual, because assets are "soft" and people can start firms with little real capital. The major predators often ignore small firms and let them enjoy early success. They let them take the risks and develop new markets.

There are an incredible number of tiny software firms. In Oregon alone we reportedly have over 700 firms, most being two people and a dog in a garage. Nationally, the average size of a software firm is under ten people. Some of these little firms do very clever things. However, at about $50 million in size, they suddenly appear on the radar screens. Very few survive that experience as independent companies. A larger, fiercer competitor snaps up their market or buys them out.

The trick is to be active in the market and gain close customer relationships, but to also be able to keep competitors from stealing your knowledge assets. You need good, intuitive market knowledge, you need innovation, you need fast cycles, and you have to be able to produce with good quality. Finally you have to be able to prevent competitors from stealing what you have, and in some cases the best way to do that is to sell out to a larger firm.

Many or most large firms are desperate for good products to fill their factories and distribution channels. Those that don't innovate well sometimes will pay a high premium. Innovative firms with compelling products, but no profits are selling for over 12 times their annual sales. In many cases multiples of 40 or 50 times sales are not uncommon. (But do be very careful. Technology theft and industrial espionage is on the rise too. Recall, the patent wars are mostly about getting cheap access to other's technology.)

The hardest part is working the gap at the R&D-customer interface. One client that is very successful in the dynamo markets calls the data collected during customer requirements interviews "murmurs." We like the term. It implies focusing intensely and for long periods to barely discernible information. In such a sea of data there are valuable nuggets of information, but it takes patient, careful listening and much expertise to convert the data to knowledge for competitive advantage.

Box 7-4

"Product development is a messy process."

Herman Miller Inc. Acceptance Speech
1997 "Outstanding Corporate Innovator" Award

Gaining understanding in complex, knowledge intensive, fast changing, high uncertainty areas is very difficult. It takes more than desire, it takes investment, skill, training, and much hard work. Often it takes luck, and even the best practitioners strike out frequently. The trick is gaining understanding before events have transpired and the answers are obvious. Late knowledge is not very helpful.

PART III
ACTIONS AND ATTITUDES

Chapter 8

The Management Challenge — What to Do?

Key Themes at a Glance

Managers trained in action, reductionist thinking, and narrow focus become very frustrated when trying to learn productive Information Age behavior. The transition from Machine Age to Information Age business is as traumatic as was the transition from Newtonian physics to quantum physics. Just as two different models of physical behavior coexist in the world of science, so too do two disparate models of organizational behavior coexist in business.

The most dangerous time is when managers schooled in old think make gestures at change. Acts like ordering downsized, overloaded, fearful groups to work harder, cheaper, and to take more risk don't help. The notion of managers prescribing and workers doing is outdated.

It is time to act. Here is what to do and what to reflect on...

- The first action is to read, practice, and learn until you personally believe new approaches are necessary. Until you develop such a deep conviction that you are willing to take career risks to drive change, you will slip back into the old behavior.

- You must unload. Effective thought rarely coexists with overload and fear.
- People matter the most, but proper metrics can help keep personalities and crisis from dominating. The trick is measuring the right things, using the right tools, and not getting lost in numbers and bureaucracy.
- Substituting Leadership and Empowerment for Management and Control is the key to winning in the Information Age. This is terrifying to control-centric Machine Age managers.
- Leaders must train in intuition and holistic thinking. These are new skills for business. As with the new physics, the observer impacts the result. Leaders often get what they expect.
- Outsight matters more than insight. Successful invention must be coupled to market need.

Back to Basics #1: Unloading

If you are a typical Western manager, you have been downsized, overloaded, and running on empty for a very long time. From the CEO down, most managers are so busy doing "it" — all the old things — which they are dangerously near burnout.

With such overload, few still have the time, energy, or resources to cope with, much less drive, change. Jobs are narrow and hours are long. Survival is the imperative and action is all consuming. Today's budgets are project budgets. Many managers' discretionary powers do not extend to a few thousand dollars of unplanned expenditures, or to hiring a consultant to help them find better ways. Some have worked 90-hour weeks for so long they can't remember anything else. Burnout and family problems are common.

A recent survey of 3,000 male readers of *Exec Magazine* showed more than half worked more than 60 hours weekly, while more than 29% worked over 70 hours. More than half have fights with their wives over this at least monthly, more than a quarter at least weekly.[1] Under such conditions, who has time to plan the future? The "vision thing" poses major problems for Machine Age managers. Most CEOs are more comfortable managing cost reduction than investing in R&D. Firms still find it easier to cling to the past than to invest in the future. Action is preferred to reflection.

Not many years ago, pundits predicted a leisure economy. They thought that the average workweek would shrink to 30 hours. In a strange way they were half-right. Many are now under or unemployed, but the rest are badly overworked. Since one group lacks time and the other lacks money, leisure time is rare and expectations are diminished.[2]

Box 8-1

"We are becoming a nation of rich slaves."

Marianne Williamson, Best Selling Author, 1997

Companies are unstable and employees are insecure. We know a manager whose job is in California. His family is in New England and he is enrolled in an MBA program in New York. A senior engineer we know has worked for four firms in three years — without changing jobs. Since the Machine Age places little value on content knowledge or translatable skills, "churning and burning" or "bungee cord management" is the trend. Some groups go through four or more managers per year.

Those trapped in the past pay a high price. A firm in a collapsing industry (defense) had a few senior executives who could see the end coming. They retained a management consultant in 1993, and his study determined they needed about $300 million in commercial-sector new business within three years. This goal was within reach, and salvation was at hand.

The consultant suggested that this would take major organizational change and a portfolio of successful ventures. He recommended training and development of a strategic framework for new business development. He suggested that each VP should sponsor a commercial initiative. As is typical with such crossovers the client had good technology, but needed to change (reengineer, if you will) most of its processes, costs, and cycle times to compete in commercial markets.

The client declined and terminated the assignment. Consumed with daily fire fighting, he couldn't afford to think about change, much less work to make it happen. As one manager said, "We think with our swords and shields in our hands." Such foxhole mentality can't help but be superficial.

In the end, the client assigned three or four engineers and a junior project manager to handle commercial diversification. Not much resulted, and their flailing and fire fighting could not hold back the future. It arrived with the predicted $300 million gap, and their business collapsed. The firm that had purchased them for $550 million in 1988 sold them for $155 million in 1994.

Machine Age managers often confuse efficiency with effectiveness. They go for cheap in the hope that it will lead to good. That seldom works. The American Management Association surveyed 830 firms that had one or more layoffs between 1987 and 1992. Only 43% wound up with higher profits and only 31% got increased productivity.[3]

Box 8-2

> **"A company surrenders today's businesses when it gets smaller faster than it gets better. A company surrenders tomorrow's businesses when it gets better without getting different."**
>
> Hamel and Prahalad, *Competing for the Future*

How might productivity drop? It drops because of low morale and the fact that finishing things is dangerous if your employer is downsizing. Remember the children's game where those who were standing when the music stopped were out? There is much anecdotal evidence that layoffs don't necessary claim the unproductive, and that the strongest are the first to leave sinking ships. Some good workers are laid-off and replaced by cheaper employees.

Those who spend time on the future are especially vulnerable. Even without "hard" academic research, employees are not stupid — they know these things. There are exceptions. Microsoft, constantly growing, trims the bottom 5% of its programmers every six months. Allen Bradley is said to have managed reductions well. Still, bureaucrats, however ineffective, are better at protecting their jobs than Information Age knowledge workers. Just consider the government.

In ancient Egypt it was standard practice to put to death all the workers who had labored on the Pharaoh's tomb, to destroy any memory of how to plunder the great wealth that would be stored there. Small wonder that tombs were seldom finished within the Pharaoh's lifetime. Two millenniums later, should managers be surprised that downsizing has had similar effects?

Anyone who ever worked in the defense sector is well educated on the hazards of being "surplused" when the job was finished. It was a common joke in the 70s and 80s that workers being laid off from completed projects would pass those being hired for new contracts in the hall.

The decline of Machine Age business has become mainstream humor. A popular newspaper cartoon, Dilbert, pokes fun at the insanity of the business world. It is set in a high tech firm where the managers are manipulative and out-of-touch, and where cynical workers toil fruitlessly at terminals in their tiny, dimly lit cubicles. The theme draws knowing nods and wry smiles, and clips can be found on the walls of many office cubicles.

The trends we discussed in previous chapters have swamped most firms. As stress increased, the typical reaction has been to run faster, cut costs, downsize, raise fear, and increase control. This is hard to escape, but it is a death spiral, which can be broken.

One insight in researching this book was — *"Managers lack the time and self-discipline to read for their own education and renewal."* That statement may sound incredible, especially to anyone who has witnessed managers struggling with overflowing in-trays, lugging piles of paper home, or experienced the endless sea of memos and reports that flood companies trying to cope with change in the Information Age.

Our early reviewer comments for this book surprised us. A few wanted learning, but most said, "Don't explain. *We don't care about why and how*. Just tell us ***what*** to do." It was these desperate pleadings that set our structure. The request itself, of course, reflects deep dysfunction. We've found that telling people what to do, without discussing why or how, is not very effective, especially for wives, clients, colleagues, and children.

We discovered that most popular business books were, in both absolute and relative terms, simplistic, theory-less, and getting more so. If *People* articles are packaged for consumption in the bathroom, most popular business books are packaged for consumption on shuttle flights — in the center seat, at midnight.

It is crucial that managers invest more in renewal and learning. World knowledge is on an exponential curve. It took from zero to 1,500 AD for human knowledge to double. By 1950 it was doubling every 50 years. By 1960 it was doubling every 10 years, and by 1992 every 1.5 years.[4]

The problem is that this learning gap is a nonlinear phenomena. Human knowledge is on a geometric curve, and it will take more than incremental efforts to keep or catch up. Letting managers burn out and constantly replacing them is not the solution.

The pace is ever faster. If the geometric progression continues, the doubling time of human knowledge goes to ***zero*** on December 22, 2012.[5] What happens then is something that is best left to philosophers. That date, incidentally, was also the last date on the Mayan calendar.

Box 8-3

> **"If knowledge is becoming antiquated at an increasing rate, we have no choice but to spend more on education. How can that not be a competitive weapon?"**
>
> Gary Tooker
> CEO, Motorola, Inc.

Adding to the problem is the fact that our educational systems are still too oriented to Machine Age training. The Dean of Harvard's Business School recently admitted that the state of business education in the U.S. was "a scandal," and for basic reasons.[6] Managers must break out and break-through.

How much "thrust" will it take to close the learning gap? Mike Kami said in 1992 that top managers should read a book a month, track four daily newspapers, and at least fifty trade journals, and take at least one three week course per year to "keep up." Kami's prescription is controversial. Some object that such "rapid reading" is superficial, saying quality is a better metric than quantity for knowledge. This group prefers deep learning and reflection. That's fine, but deep learning and philosophical thought takes even more time than Kami's recommendation. Charles Handy suggested in 1994 that a minimum of one day a week be devoted to learning and self-renewal.[7] He personally spends more than half his time thinking and writing.

How can managers move from the present, where essentially no time is invested in self-renewal, to 20% or more? Obviously, routine work must be severely reduced. You can't live in both the Machine Age and the Information Age; you must choose your path. As Handy says, "In the end there will be no results if there is no investment in people."

Warren Bennis, the management Guru, reports that his moment of truth came at 4 am, after ten months as President of the University of Cincinnati.[8] He was still trying to get through the day's mass of paper. He looked back through his calendar, wondering where all his time had gone. He discovered that he was "the victim of a vast, amorphous, unwitting, unconscious conspiracy to prevent (change) in the status quo." Bennis came up with a law, "Routine work drives out all non-routine work ...smothering all creative planning and fundamental change."

The solution? Unload. Do less. Think more. Invest in self-renewal. Solicit outside-the-box viewpoints. Gain *deeper* understanding. Learn how and why. Take *informed* action. Target *causes*, not symptoms.

Box 8-4

> **"I have gotten very busy. Today I have so much to do that I will spend two hours meditating instead of one."**
>
> Mohandas K. Gandhi

Start by eliminating at least half of your workload. Drop the least important 20%. For the rest, don't delegate; transfer ownership. You must unload if you are to change. If you mix the urgent with the strategic, the urgent will always dominate.

When should you start? We suggest that the best time is ***now***. Certainly, the leaders, the top 10%, are already using Information Age methods. Even in the next tier, the "rapid followers" are trying to change their Machine Age firms into more effective cultures and structures, though without much success.

We say start dealing with the future *now*, but many firms are too battered, downsized, and frightened to act. We don't hold much hope for these, but a few may still have one last chance to break out. Dr. Charles Handy had a thoughtful suggestion. He calls it the Sigmoid curve. See Figure 8-1. Handy suggested that point A, the midst of a recovery, is the best time to start new businesses in *The Age of Paradox*.[9] You should light candles and view the spurt of revenue from an economic upturn as your last chance for renewal.

Resource scarcity is less of a problem. Handy notes that very soon half of the workforce will be "outside" the company. That suggests how firms can staff for planning, diversification, and upswings. They can use contract help and consultants to assure results before the next downswing. If you can't act now or during the next upturn, start planning to sell the company.

Back to Basics #2: Finance

We noted earlier (Chapter 3) that the traditional Machine Age financial methods were sometimes inappropriate for Information Age business,

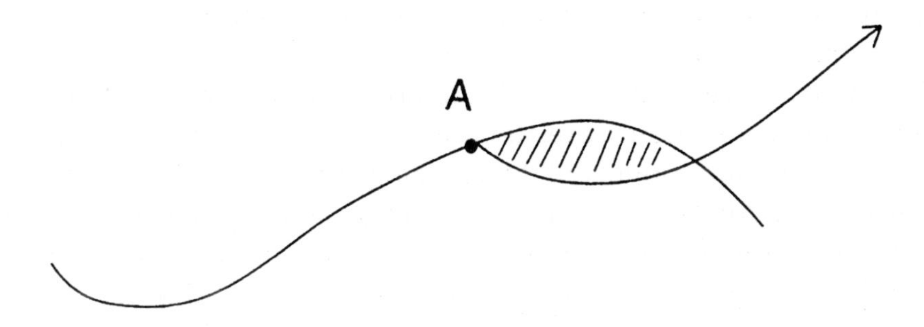

The sigmoid curve describes a pattern of growth and decline. Investment at point A starts a new curve (renewal). At point B, investment is massive and significant.

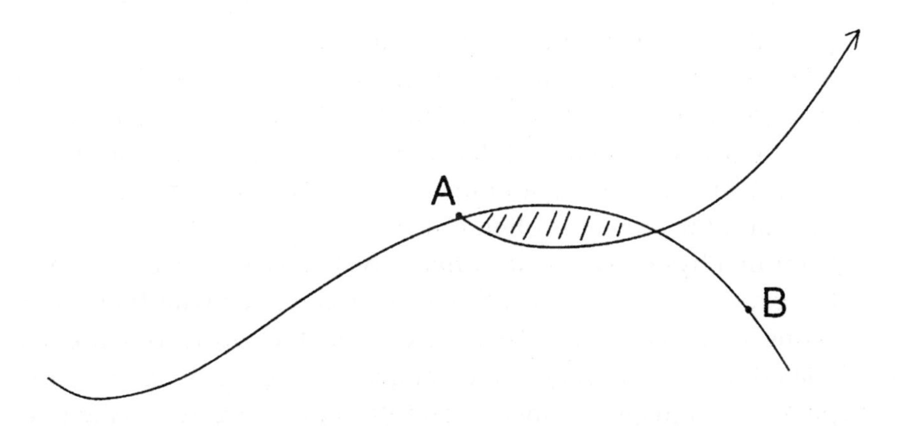

Figure 8-1: The Sigmoid Curve Applied To Renewal. Adopted from Charles Handy, *The Age Of Paradox*. Boston, Mass.: Harvard Business School Press, 1994, pg. 52.

and gave examples. One group always asks, "So what? Those are the rules." These see business as finance, for that is how they were trained. Some are harsh, saying that without numbers nothing can be trusted. Others think it wastes time to ponder things which "can't be changed." These people are trapped in a self-imposed prison.

Traditional finance interferes with Information Age organizational communications and business relationships. Accounting is the language of business communication, but it was a language designed primarily for the Machine Age. If we want to create a new and different world of business we must combat the bean counter mentality. Change is impossible without addressing the issues of linear thinking and financial abstractions.

MIT has long been a leader in the science of learning organizations. They have a professor of Managerial Accounting, Fred Kofman, who is a student of the philosophy of language. He has come up with models that he calls "double-loop accounting."[10] The term comes from communications research, where experts have shown that overly concise communications (e.g., bullet charts and executive summaries) can get in the way of useful understanding.

Most associate accounting with scorekeeping and calculation, not with communication or language. In the Machine Age, the age of Newtonian Physics, that meant that after some business has happened the accountants came along and measured the results. That is the old meaning, and the one most often practiced.

Quantum Physics teaches us differently. It tells us that the observer and the experiment interact intimately. It is just as true that when discussing business issues the speaker and the world outside are interrelated. While a speaker creates language, language is also creating the speaker. Language is a mechanism through which we create new realities and understandings as we begin to explore and talk about things of interest. Traditional accounting can limit communication.

Think of what "to account" means. At the core, is it not more than a simple financial calculation? Is not the goal to explain something's purpose and history? Doesn't "to account" include values as well as numbers, feelings as well as facts?

Kofman's double-loop accounting provides a meaningful dialog to help people better share assumptions and gain understanding. Proper attitude is crucial. Instead of sharing divine truth, the presenter is asking for help and involvement. The audience reacts not with attack, but with inquiry.

A typical dialog might be, "I am struggling with this. What do you think makes sense in this situation? Under these assumptions, we get the following financials." The presenter exposes all the assumptions and alternatives of which he is aware. The audience might react with something like, "Oh. You lost me when you jumped to line twenty-three, which shows a return of 20%. Couldn't we do better if we outsourced?" The reply might be, "I don't know. Outsourcing makes me very uncomfortable, because it means shutting down one of our plants."

Some relationship experts tell us that there are five levels of human communication: cliché, fact, opinion, feelings, and union.[11] The deeper the level of communication the greater the value and the higher the risk borne by the speaker. It takes self-confidence and lack of fear to work deeper than the fact level.

Just as few couples have the intimacy and trust needed for deep level discussions, few firms have the trust and openness to feel comfortable with the above dialog. It is safer to pretend the numbers are facts. But they are not facts. *There is no objective reality, only what we create ourselves by our interactions with others and with events.* Information Age firms are learning to communicate at deeper levels and with new languages. That is a fact, and the leaders are winning because of it.

Back to Basics #3: Metrics

In Chapter 3, we noted that the traditional metrics of business were not very helpful in the Information Age. That does not mean that we are opposed to metrics. To the contrary, winning in the Information Age requires timely knowledge.[12] If it can be quantified, so much the better.

The best firms are very good at measuring the things that matter. Modern technology allows the vital signs of a business to be measured in real time, and made available to managers and to workers. The important thing is to understand *why* metrics should be used. The main reason is to move the important business decisions from traditional dominance of "the last comment received" and "the most persuasive (or loudest or most assertive) manager" to more rational and scientific criteria. The main purpose is to guide investments, better sense the environment, and allow continual improvement. If the data collected does not allow effective high-value action, don't bother collecting it. If it does, invest lavishly.

Manufacturing operations are the easiest to control. It would be impossible to build modern semiconductors without statistical process monitoring and control. These methods are well known thanks to Doctors Deming, Juran, Crosby and many others. There are many good books on the subject.[13]

The Software Engineering Institute has a well known five-level "maturity model," as shown in Figure 8-2. Colleagues report that the average U.S. manufacturing operation is at level 4. The average engineering organization is only at level 2.[14, 15] (Note: The SEI work came out of the Department of Defense. It is a quality centric metrics-based approach to "big-program" software development. Internet firms, and Microsoft, use different approaches that favor experimental activities — like extensive Beta-testing — over metrics.)

It is entirely possible to monitor and improve engineering effectiveness if you view design as a process and let people stay in the same jobs long enough to improve. The trick is to measure a small set of parameters, and to set up "feedback loops" (including training) to reinforce the desired behavior. The key is to do this in an objective, non-personal, and statistical manner, with a focus on improvement rather than punishment.

Again, it is crucial to measure only a small number of things, and to separate personnel issues and blame from the system. *If fear exists, metrics based systems seldom work.*

The following characterizations of the five maturity levels highlight the primary process changes made at each level.

(1) *Initial*	The software process is characterized as ad hoc, and occasionally even chaotic. Few processes are defined, and success depends on individual effort.
(2) *Repeatable*	Basic project management processes are established to track cost, schedule, and functionality. The necessary process discipline is in place to repeat earlier successes on projects with similar applications.
(3) *Defined*	The software process for both management and engineering activities is documented, standardized, and integrated into an organization-wide software process. All projects use a documented and approved version of the organization's process for developing and maintaining software.
(4) *Managed*	Detailed measures of the software process and product quality are collected. Both the software process and products are quantitatively understood and controlled using detailed measures.
(5) *Optimizing*	Continuous process improvement is enabled by quantitative feedback from the process and from testing innovative ideas and technologies.

Figure 8-2: SEI Maturity Model.

Box 8-5

"Metrics are a double edged sword, representing almost as much potential hazard as benefit to the user."

Marvin L. Patterson
Accelerating Innovation

Machine Age firms often misuse metrics. One classic is the apocalyptic tale of a control centric upper manager who noted that his firm's products were aging. He ordered that his general managers' compensation would henceforth index to the percentage of their sales coming from "new" products.

The result, of course, was exactly opposite to that intended, and the consequences severely damaged his firm. The managers being threatened shifted their businesses to brand new old products, by making minor changes or re-nomenclating. Things like this have happened so often that heads nod when we mention it in lectures.[16]

Marvin Patterson related a wonderful story of misused metrics to us. One firm decided to rate their secretaries by how many keystrokes they produced. A program was implemented, and data were collected. The number of keystrokes "improved" dramatically. Machine Age control was working, or so it seemed.

The reason was finally discovered when a manager observed a group of secretaries talking over lunch. Each had her thumb firmly pressed on the space bar of her terminal.

With fear banished, proper metrics can help. A former V.P. of Technology at Tektronix, Dick Knight, related a story of how their partner, Sony, does this.

Sony and Tek have long had a joint venture that sells Tek products in Japan. He was surprised that a visiting Sony-Tek marketing manager had statistical data on how long each Tek customer support person took to return calls. When Dick wondered who the good and bad performers were, the Sony-Tek person (a Japanese) immediately closed his notebook and changed the subject.[17]

How might this guide Sony's management action? If something was urgent or crucial, they would know they should call person X. For the routine or unimportant they would not bother person X, but would call person Y instead and accept the slow response.

Sony's goal was knowledge, not punishment. They know that Western systems are blame-based and that fearful people distort metrics, so their manager changed the subject when challenged. This was probably done to protect the system, not the people, but the people are part of the system. "Deal with problems, not with people," is very Japanese.

The relevant concepts are learning organizations, statistical process control, and systems understanding. If metrics are collected for training and improvement instead of for punishment and rating, they can help. The goal is to manage the things that lie beyond one's direct control.

Image and brand awareness in another area that lends itself to metrics. Since Public Relations people maintain that "perception is all there is," it seems fitting that Intel collects metrics to manage their worldwide Public Relations effectiveness at a detail level.[18]

All over the world, each meaningful media message about Intel is clipped, transcribed, and rated for its tone and contents every month. Mentions without content are discarded. Messages of substance are graded for content (positive, negative, or neutral) and tone (favorable, unfavorable, or neutral). Statistics are collected at a detail level and reported to top management at least monthly, with an executive summary. These metrics are used to guide action.

How might this help guide Intel's management action? They will know, by product if necessary, what key messages are being received and how they are being perceived. That helps them better manage their advertising, sales, and PR programs. For example if some message they are sending is being negatively reported, they can take corrective action. (One wonders what tremors rippled through their system during the famous Pentium bug episode.)

They can know, by individual, which of their key executives are being quoted, and which are credible to the press. If some executive is often quoted, but the quotations are neutral or negative, they may conclude that he is overexposed and shelter him for a time. If one is seldom quoted, but always positively, they may increase his exposure.

The presenter, an Ex-Intel PR person, was emphatic that what was being measured was external events — not PR people or PR programs. He said, "We do programs. Events happen."

He alluded to the bumper sticker, "S___ happens." Indeed. For example, press coverage in Taiwan turned very negative for a period. The reason had nothing to do with Intel's PR, but with their decision to assert a patent.

The same methods can be applied to monitor and manage other aspects of the "outside world," in this case, the press, columnists, and writers. They will know which individuals and which publications are most inclined to report certain things, and how they might be inclined to report them. Using such deep knowledge intelligently can increase effectiveness.

Microsoft usually gets favorable press, and it always has. Can this be just luck or the skill of their PR people? We doubt it. Microsoft is a smart, tough Information Age competitor. Might they not monitor and manage their PR as well as or better than Intel? If only they bothered to do as well with their industry relationships, which might yet be Bill Gate's downfall.

Box 8-6

"As the Microsofties gaze from their castle walls at night, that's not the glow of dawn they see in the distance. That swelling light comes from the torches of the peasants. And the mob is made up of people it has dismissed as unimportant all these years, as unworthy of attention and courtesy."

Jesse Berse, Editorial Director, Windows Watcher Newsletter

Intel and Microsoft are, of course, wealthy and sophisticated firms. If you can only measure a few things, what might be most important? Here are some suggestions:

- Cash Flow
- Sales calls to targeted customers
- Customer service rating of each person
- Sales and orders (they are <u>not</u> the same)
- Quality
- Progress against agreed to goals (frequently)
- R&D effectiveness (innovation)

Back to Basics #4: Empowerment

Empowerment is a vastly different thing than delegation. Without leadership there can be no empowerment, and there must be many levels of leadership. When trying to build self-managing organizations, the Generals should be leaders, but so should be the Captains and Lieutenants.

When Kouzes or Posner lecture, they often use the "Snoopy" cartoon shown in Figure 8-3. Snoopy the dog is leading a Boy Scout-like troop of birds. He decides he needs a flag for them to rally around. He chooses a flag with his picture, but, after discussion, they find they each need a flag with their own picture. Each needs his own vision.

Many managers reject shared vision and empowerment as "too slow." That's wrong. Most gurus of organizational change say that it takes a long time to change a large organization. That's wrong too. Meaningful change through empowerment can happen in hours.

In one reported case, the leader of a deeply troubled and about-to-go-out-of-business organization turned it around in *one day*. The organization consisted of 7,500 people and it had a history of decades of Old Think, Machine Age management.[19] A single twelve-hour meeting changed the whole company and let it survive.

The firm made nuclear reactors for submarines, and had suffered from low morale, poor quality, and even sabotage. This is a very unforgiving market. The best case of a major failure would lead to the loss of the ship and the death of the crew. The worst case would cause a major ecological disaster. The first step to driving major change is getting people to know in their hearts that it is both necessary and safer than continuing with the status quo. In this case, the leader used the sabotaged equipment and the horror of the possible consequences to convince his managers and workers that immediate change was mandatory.

The way of empowerment is not complicated, but it is very difficult. It takes courage, and willingness to change. Behind and following that 12 hour meeting was a CEO who supported the President, a President who led and empowered, a superior team of mentors and coaches, and months of assiduous work by many people. Of course, the fact that management was honest and that everyone believed at the core of his or her being that their continued employment depended on major change was no small factor.

Fig. 8-3 PEANUTS © UNITED FEATURE SYNDICATE.
REPRINTED WITH PERMISSION.

Step one: Leaders must always challenge the process. The Leader must develop a clear and compelling vision. The vision must be able to be shared and adapted by all those involved. The vision must be noble and inspiring.

Selfish visions — e.g., I want to make a lot of money — don't help. Who wants to work to add to the wealth of a power boss? We have never seen a financially centric vision provide much useful guidance. At best financial based visions are not very noble. At worst they are cynical, "We abuse our employees and pass the savings on to you."

A vision to become good at something of value to customers (be it microprocessors, wireless phones, fried chicken, or athletic shoes) can lead to limitless profits. Conversely, a vision of being good at profits is usually hollow and empty. As we said before, the traditional financial metrics don't help to sort healthy vital companies from those who are profitably going out of business. Since the latter case is much easier to manage, why be surprised when it results? You asked for numbers... and you got them.

Useful visions for corporations seem to almost always contain words like "best" and "excellent." Best at what? Best where? And best to whom? Useful visions for social organizations seem to often contain concepts of "shared dreams." Consider the famous Martin Luther King "I have a dream" speech, or John Kennedy's vision to put man on the moon. These were well-crafted visions. They left room for individual accomplishment, but gave clear guidance to all members of the team.

The key point is that there are *many* visions that can lead to prosperity. Every person and every organization has a right to a dream and the ability to be the best at something. There is not one future, but thousands. What distinguishes greatness and what marks leaders is the ability to uniquely imagine what can be and then make it happen.

From art or acting to music or sculpture, there are many successful artists, each with different styles and different visions. The success of one does not ordain the failure of another. Still the distinction between greatness and mediocrity is usually clear in the end. Is it not the same in business?

Step Two: Each work group must decide what "great performance" means in the context of the guiding vision.

General Norman Schwartzkopf had a vision of quickly destroying Iraq's ability to wage war by a new form of battle called AirLand.[20] He envisioned using smart weapons and smart troops to conquer a superior defending force with few casualties.

The vision was radical. Conventional wisdom said that this was impossible, even absurd, but the vision was compelling. Over months, this was translated into a vision for the Air Force (e.g., "Destroy the air defenses, communications links, and radars, by..."), for the Army (e.g., "Make the enemy tanks move and kill them from the air, by..."), and so forth.

Each General then developed a supporting vision of how to win his battles. Each Colonel developed a supporting vision of where and how to attack. This process continued down to the lowest levels of independent action.[21] As with Snoopy, each leader developed his own vision.

Step Three: Each Group identifies the obstacles preventing great performance. They identify who owns the obstacles. (Ownership has two levels, the performer for performance and the manager for empowering.) They ask what actions the owners will take to remove the obstacles.

Box 8-7

"Leadership means removing the barriers that prevent people from taking pride in their work"

W. Edwards Deming (repeated often in his lectures)

At the top level, Colin Powell owned the main obstacle — Washington DC, including Congress and the White House. The main risk was that of interference or impulsive decision. On many occasions, Washington's ignorance and lag time could have commanded action that was premature, belated, or otherwise dysfunctional. In some minor areas, this occurred.

In the main, though, Powell did his job and that let Schwartzkopf do his job. That job was setting realistic goals, planning how to accomplish them, and making it happen. He did this by developing a vision, a battle plan if you will, and then sharing it with the people who had to make it happen. Each level took Schwartzkopf's vision, and blended it with their own for operational purposes.

There were non-operational issues. Fighting a war in fast cycle compressed time necessitated Information Age infrastructure and methods. The only battlefield promotion given was to General Gus Pagonis. He commanded no combat troops and fought no battles. He managed the logistics for the war, and he succeeded in removing the major obstacles to performance that faced the fighting forces. Schwartzkopf felt a key to winning the war was giving Pagonis an equal rank to the combat commanders, so he could do his behind-the-scenes job without interference.

Empowerment works. Rather than drag on for years, as in Vietnam — run without vision, but with Machine Age control and micro-management — the Gulf war was decisively won in 100 hours of combat. Cost and casualties were minimal. In similar manner, breakthrough business performance can be achieved.

The most important thing to remember is that in the Gulf war, the performers and managers were left alone for a relatively long time, over six months. They needed this time to develop multiple levels of shared vision, and to work their way thorough all the obstacles. They needed to practice as teams, to make mistakes, and to learn. Had they not been given this buffer, Information Age leadership would have failed and the Gulf war might still be in process.

The second most important thing to remember is that all this looks easy only in retrospect. The key army commander whose job was to assault Hussein's main force, General Fred Franks, was chronically apprehensive and late. Is it any wonder, given that his mission was to charge headlong into a much larger force of elite Republican Guards that had the advantage of defending fortified positions?

General Franks had to trust his leader and trust that others would guard his flanks and disrupt the enemy's ability to resist. He knew that if the other members of Schwartzkopf's team faltered, his mission was suicide. His troops would be annihilated and his career shattered. Schwartzkopf spent much effort and attention reassuring Franks, making him know that his role was crucial, and urging him to attack quickly.

Schwartzkopf was deeply concerned that Franks would arrive too late to destroy Iraq's main force, which would escape to fight again. Franks was worried that his task was impossible without more forces. As it turned out the vision was sound, the team all did their jobs, and Iraq's resistance collapsed before Franks' assault. Instead of "the mother of all battles," it was a major rout and allied casualties were minimal.

What made the difference? Leadership and empowerment. The people responsible for winning the war were empowered, given what they needed, and made accountable for results. Rather than being told how to do their jobs, the local commanders themselves decided how the jobs could best be done. Then they did it, both individually and as an interdependent team.

Unfortunately, empowerment or trust in either business or warfare is usually rare and often fleeting. Backsliding is common and leaders seldom hang around to become bureaucrats and administrators. What we have witnessed since in Bosnia or Somalia or Haiti looks more like Vietnam or Korea than it does the Gulf War.

Steps Four and Five: In addition to the actions described, researchers have noted that leaders also model the way and encourage the heart.[22]

We are over managed and under led. The Machine Age models of workers and bosses do not extend to vision and empowerment, much less to discussions of feelings or to training by example. Most management books speak of control, not trust, empowerment, or shared objectives. This will change. Our society in general, and business in particular, needs to learn more about meaningful human relationships to again enjoy happiness and prosperity.

Box 8-8

> **"We have to distrust each other. It is our only defense against betrayal."**
>
> Tennessee Williams
> U.S. Novelist and Dramatist

Business need not be grim routine, it can be fun. Trust-based business relationships are possible, and, in fact, common in Asia.[23] Empowerment can be taught, learned, and practiced, but first you must clearly see the benefits and want to change badly enough to leave your comfort zone and actually "just do it." This is like the old joke, "How many psychiatrists does it take to change a light bulb? Only one, but first it must want to change."

Back to Basics #5: Intuition

Most of Western training is based on reductionism, the idea that things can be best understood by breaking them into small pieces. This is the core of Machine Age thinking or Newtonian physics.

Box 8-9

> **re·duc·tion·ism (rî-dùk¹she-nîz´em) noun: An attempt or a tendency to explain complex phenomena or structures by relatively simple principles, as by asserting that life processes or mental acts are instances of chemical and physical laws.**
>
> The American Heritage Dictionary of the English Language

For example, the cars of the 60s could be understood by taking them apart, looking at the gears and parts, and then putting them back together. That approach never worked well in areas like software, and this is where the gaps in Machine Age management first showed up. Books like *The Mythical Man Month* pointed out that while adding labor might speed ditch digging, it often had the reverse effect for software development.

There are more opportunities than ever for specialized products, but today's markets are very unforgiving to products and companies that fall short. The cars of the 90s have evolved into inter-related systems composed of electronics and software. Reductionism now fails even there. Honda's technology, quality, and systems integration, beat Volkswagen's functional simplicity.

The opposite of reductionism is Zen or holistic thinking. Some in the West call this a strategic attitude. Rather than breaking things into tiny pieces, the practitioner attempts to gain mastery, and to deeply understand both the whole and the minutia of detail that create it.

Frederick Betz writes on the management of technology. He thinks that the reason for such a gap between Eastern and Western approaches is that in the West the literate caste were often priests, while the warrior caste was usually illiterate.[24] In the East the warriors were highly educated, along with the priests. Perhaps this explains why we tolerate such a high level of illiteracy (about 40%) in the U.S. workforce.

Betz draws two core conclusions, both of which surround the fuzzy word "intuition." He concludes that there must be a role for the experience that only action provides.

- Perception, commitment, and competency are the fundamental components of a strategic attitude.
- One cannot be intuitive if one lacks experience.

We suggest that this is fertile ground for management thinking, and certainly this is the approach preferred in Pacific Rim and by some leading U.S. firms. If you prefer the Eastern approach, you should read works like Miyamoto Musashi's *A Book of Five Rings*.[25] Musashi was a man of action. He learned through doing, and he fought 60 sword duels without a single defeat before turning to writing. Betz paraphrases Musashi's way for Samurai warriors in Western management terms as:

1. Do not think dishonestly.
2. The WAY is in training.

3. Become acquainted with every art.
4. Know the ways of all professions.
5. Distinguish between gain and loss in all worldly matters.
6. Develop intuitive understanding and judgment for everything.
7. Perceive things that cannot be seen.
8. Pay attention even to trifles.
9. Do nothing that is of no use.

The most interesting thing about these rules, which we deem appropriate, is ***why*** they apply. The reason is not morality, but effectiveness. Consider rule one: If one thinks dishonestly, one deceives one's self. This prevents clear thinking and useful vision.

Business has a problem with rule two, which largely negates the basis of Western education. Managers like to think that taking classes is adequate, the shorter and cheaper the better. They find it hard to accept that newly minted MBAs, who paid enough for their degrees to buy small houses, are untrained because they lack experience. Few bother to groom subordinates by letting them make mistakes and coaching them. In fact, seasoned employees are often seen as more expensive, hence disposable. All sports teams practice, but most business teams do not.

Why one should go to all this trouble and work to learn new attitudes and skills, especially when we have managed to avoid it for so long? The core assumption is that all action (rules three and four) occurs within a complex, chaotic, holistic context — a system in which many skills are needed to carry out effective action. Reductionist, linear procedures just do not work in such a context. Is this is not as true for knowledge-based business as it is in warfare?

In such an environment, the traditional Western separation of planning and action is hopeless. This makes intuitive facility in the heat of action crucial (rules six and seven). All military leaders hone their intuition, but most business managers do not.

Anyone who has managed software or Pentium chips knows the value of rule eight. Small bugs have cost many firms dearly. Rule nine acknowledges that key resources are always scarce.

If you prefer not to learn from the Japanese (a mistake), there are other models. Jack Welch of General Electric uses six similar rules, which he calls strategic attitude:

- Face reality as it is, not as it was or as you wish it to be.
- Be candid with everyone.
- Don't manage, lead.
- Change before you have to.
- If you don't have a competitive advantage, don't compete.
- Control your own destiny, or someone else will.

Note the common focal points of these behavioral rules, developed centuries and half a world apart. Scientific research repeatedly has shown that the #1 characteristic of superior leaders is *honesty*. Kouzes and Posner surveyed 2,615 U.S. managers, and 83% of them deemed this as ***the*** most crucial leadership trait. Here too there is a gap. Only 32% of the public believes most corporate executives are honest — 55% think most are dishonest.[26]

All seasoned consultants speak of how difficult it has become to qualify clients. Disconnects are often caused because different levels of management are not being honest and open with each other. It is sad to work for months developing understanding, empathy, and rapport with a client, only to find out that the needed work is blocked.

Few situations have that much drama, though a venture capitalist we know relates a tale where the CEO and the President got into a fist fight at a board meeting, knocking over chairs and endangering others in the small room. More common is the random reorganization, which resets everything in the company. This happens frequently.

Once when a prospect was severely downsized, we called seven separate contacts before we found a person who was still there. Naturally, the lone survivor, witnessing the carnage around him, had little interest in doing anything until the dust settled.

Another too-common case is working closely with a client to mutually define and agree to a high value assignment, only to have the work vetoed by someone higher up. The person who "managed" the business turns out to have lacked the job content necessary to do the very thing he deemed most important!

In one case the COO of a privately held firm defined a major assignment to move his fast growing company from a custom system integrator to standard products. With inexperienced, junior help he was frantically spinning dissimilar plates on poles spread all over the planet. He desperately sought standardization, and we devised a plan. But the CEO vetoed the assignment, preferring long hours and cheap workers to investment. He was happy with the job shop environment, and if his COO could not keep up he could hire another. Why build fire trucks if buckets work?

In another case, a Fortune 50 company set up a diversification business in an emerging market area they knew little about. They acknowledged this ignorance by literally walling the business off from the rest of the firm, providing a separate budget, and autonomous management. But when the General Manager of the new venture wanted to spend a small part of his own budget for consulting services to help sort out opportunities, corporate vetoed the contract. They had been laying off and had surplus people they wanted him to use.

Back to Basics #6: Outsight

One thought that needs to be clearly grasped and believed, is that the most important things lie outside your firm. Nothing is more dysfunctional than traditional Machine Age NIH ("Not Invented Here") rejection of outside ideas. The most certain way to ensure business failure is to close yourself off from outside knowledge and "outside-the-box" viewpoints.

Intellectually most people know this. James Utterback of MIT is one of the country's leading authorities on innovation. He says, "Market forces seem to be the primary influence on innovation. From 60 to 80 percent of important innovations in a large number of fields have been in response to market demands and needs."[27] Many others have come to similar conclusions, as have we.

Folklore has us believe that small firms gain a freedom from bureaucracy that renews their ability to innovate internally. That is only half-right. Most of the small firms we have come in contact with, including those we have helped start, used innovations and market knowledge from outside the firm. The stereotypical example is the group of founders who grew frustrated that their employer would not act, and so left to form a successful start-up.

Some recent research indicates small, high tech electronics firms had their best successes when they avoided risk, stuck to things they know well, and sourced external technology, and perhaps distribution.[28] If you are cash poor, but agile and have low overhead, you can sometimes steal your livelihood with little risk from larger predators.

The study was done in Huntsville, Alabama. One gets the picture of laid-off defense workers doing what they know best and gleefully applying it to commercial markets, and of large Machine Age firms unable to react because of ponderous procedures, high costs, and NIH syndrome. The key point is that the source of innovation was external to the firms doing it.

This is often the case. Bill Gates did not write DOS, he licensed it. The Japanese did not invent semiconductors or flat panels, they licensed them to get started. Apple did not invent the Graphical User Interface (GUI), they got it from Xerox PARC, as did Microsoft when they did Windows. Neither Compaq nor Dell nor Packard Bell nor HP invented the PC, but they all had good market shares at this writing. (IBM, who did invent it, limps along in fourth or fifth place.)

It has always been that way. RCA did not invent television, and neither did the Japanese. Boeing did not invent the jet airliner and Henry Ford did not invent the automobile, but all these firms made a lot more money than those who did invent these things.

If there is one myth that needs to be buried and forgotten, it is the "better mousetrap" notion. The person who said that inventing a better mousetrap made the world beat a path to your door was a poet, not a businessman. Business just does not work that way.

Thomas Edison worked backwards from market need to invent electric lights. Electricity was more costly and less reliable than gas at the time, but it was the "killer app" of the day. If you used gaslights and something went wrong, your home blew up or caught fire.

Perhaps because of the better mousetrap notion, there is something relentlessly myopic about the U.S. approach to technology and business. It shows up in government. Spending public money on technology creation is apparently good, but spending money on commercial application of technology violates dogma.

When we were spending $70 billion a year on cold war technology, the government insisted — and industry, the press, and the voters apparently accepted — that the U.S. had no technology policy.[29] Sadly, this lavish funding has created a defense ghetto, where defense firms create expensive systems from archaic technology. The Patriot missile, which got so much favorable publicity during the gulf war, did not even use microprocessors. Its technology was vintage 1970, and was several generations obsolete by commercial standards.

As with the apocalyptic *Japan that can say "No,"* you can't buy anything but sanitized and watered down versions of Fumio Kodama's writings in the West.[30] Kodama's core concept is simple but revolutionary. The old breakthrough models of innovation are being replaced by a more dynamic system. He calls this "technology fusion" — the combining of already existing technologies or engineering disciplines into new hybrids that are greater than the sum of their parts. Merge electronics and mechanics and you get mechatronics.[31] Merge electronics and optics and you get opto-electronics.

Under this model the winning firms don't create technologies, they rapidly absorb them — in Kodama's words, "Like waves crashing onto the beach." Thus Canon moved from cameras into mainstream office equipment. Thus NEC moved beyond public telecommunications into computers, semiconductors, and wireless phones. Thus Fujitsu spun off Fanuc, one of the most profitable firms in Japan, to make mechanical robots. Thus the Japanese, with Sharp in the lead, gained control of the huge and rapidly growing world market for flat panel displays.

Kodama's conclusion: the critical resource for competitive manufacturing is not capital and labor, but R&D. By that he means the new type of R&D, not the old technology creation type.

Japan is using exactly the opposite strategy to the cost containment and denominator management of Western Firms. The ratio of R&D to capital expenditures crossed over in 1985 and was 1.26:1 by 1987. By the 90s, metrics had been refined to measure this more closely. By 1995 their high tech firms were spending an average of 80% more on R&D than on plant and equipment.[32]

Think of this as "demand side" as opposed to "supply side" R&D, a discipline of focus on outsight as opposed to insight. This is a serious scientific art, not a casual flash from some upper manager. Doing competitive products in today's globally competitive requires skill, training, and corporate focus, not knee jerk reactions to the last request a customer made to a salesperson or upper manager. Understanding requirements deeply takes so much more than surveys or focus groups.

Myth: Most companies know what their customers want.
Facts:

- What companies think they know about customers is more likely to be wrong than right.
- Customers rarely buy what companies think they sell them.
- Customers in reality define the business. What the customers think they are buying determine value and quality, not internal metrics or what management thinks.

One good example of this was Intel's 1994 decision to delay fixing the Pentium bug. The decisions to push forward the Pentium's introduction and to stonewall its bug problem involved so much money they became upper management prerogatives.

At a technical level, the bug was trivial. To an engineer — or to an upper manager with a rational, technical background — it was much less serious than bugs in previous processors or in mainstream software. Perhaps customers "should not" have cared about such an esoteric bug, but they became infuriated when Andy Grove told them so. The louder and more precisely he told them, the angrier they got. The longer he delayed a recall, the worse the problem grew and the more expensive the solution became. In the end it cost Intel almost half a billion dollars, plus considerable ill will, and a public pledge to reveal future bugs when discovered.

What had changed between previous chips and the Pentium? The "Intel Inside" campaign made that chip something that its predecessors were not — a branded consumer product. To the customer, Intel, not the PC vendor, was the culprit. They viewed the problem like taking a chance, however slim, of buying contaminated drugs or a car whose brakes might malfunction.

That the odds were low of any given customer being effected by the problem did not matter. It would not have mattered if the odds were one in a billion, because it was a perceptual or a marketing issue, not a technical issue. The product was defective, and customers wanted it fixed. In such issues, as Mr. Grove discovered, the customer is always right. Fortunately, Intel is very good at new products, so it will probably learn from the experience and recover.[33]

To Intel's credit, the correct decision — product recall — was reached after a few weeks of public agony. The interesting point to ponder is the tension between top management control and local empowerment. What would your firm do in a case if 1) those close to the product were advocating customer accommodations, 2) the accommodations would disrupt your firm and cost hundreds of millions of dollars, or more?

As the world speeds up we can expect that such cases, where local product initiative and corporate strategic issues overlap, will occur more often. What processes does your firm have to resolve such conflicts? At what level must such decisions be made? How long would it take those at that level to become fully informed and make a decision? We suggest you will find it less painful to learn these answers before a crisis develops.

This issue is being called "envelope supervision." The key question is how might upper management balance their responsibility for controlling the firm with the need to empower for local innovation?"

This needs more attention.[34] There are, incidentally, much worse scenarios than the Intel case. A young, fast moving and empowered financial trader recently made an error that cost close to a billion dollars, causing the bankruptcy of the two-centuries-old investment bank, Barings that employed him.

For most firms, the attempt to overlay market "outsight" on top of management control has not been very successful. This has reached crisis proportions in the area of new product development. The lifeblood of firms comes from new products, but upper management intervention, albeit in the name of the customer, has probably done more harm than good. Because of management overload and other reasons, most efforts are sporadic and superficial.

Shifting from the insight of corporate tradition and better mousetraps to the outsight of market responsiveness is harder than it seems. The part of new product development where senior management can most help is at the fuzzy front end. Unfortunately, few managers are comfortable working in an environment that is vague, risky, and difficult. Decisions made here have largely unknown long-term effects, and are almost always imperfect to some extent.

Most high technology firms develop crisp systems for their new product development (NPD) processes. Unfortunately, if one diagrammed what was actually practiced, the results would be totally bizarre. The mere act of honestly diagramming "as is" process is so threatening that it has resulted in more than one consultant being fired.

Box 8-10

"This diagram (of actual NPD practices), that looked like a cross between something drawn by Ronald Searle and Rube Goldberg, depicted a process so chaotic that it is almost impossible to describe it verbally. Among other things, it consisted of grotesque bulges arising from the late infusion of new ideas into the process, strange sidetracks, and unwarranted filters, most of which reflected the effect of senior management's unwarranted interference in an essentially orderly process."

Laurence P. Feldman, *PDMA Visions*, January 1995, pg. 16.[35]

Upper management can help, but not by late intervention. A better use of upper management time would be:

- Creation of trustworthy, seasoned, multi-disciplined cadres (perhaps including consultants and contract help) to specialize in NPD.
- Definition of project charters, linking success to corporate strategy and business results.
- Provision of continual upper management sponsorship, support, feedback, and guidance through the NPD cycle.

A few years ago one of the authors participated on a panel of experts in NPD at a major Silicon Valley Function. The panelists had not met before, but all were experts in the field. We were amazed that we all totally agreed on several key issues.

- Most firms made their worst NPD errors in the concept phase, and most firms seriously under-invested in this phase.
- The greatest percentage of new product failures resulted from missing the market.
- The concept phase should take about half the calendar time and 15–30% of the project budget. Deliverables should be a plan, budget, and schedule, and a "chicken test" consisting of an actual demonstration unit.
- Upper management owes the NPD team and the stockholders a prompt Go or No Go decision at the end of the concept phase.

    * This (or earlier) is by far the best time to kill projects.
    * The "stutter start," where projects are turned on and off, up and down, almost always leads to late products, missed markets, low morale, and other bad outcomes. Fiscal rather than project budgeting virtually assures this will happen.
    * Project approval without the actual timely commitment of agreed to resources, is common. It causes late products, lowered financial returns, and other bad outcomes.

- Most firms would be better off to do fewer products, but better and faster.

In summary, managers need to become more outward looking, but they also need to keep in mind that there are many appropriate time horizons and levels of outsight. All need better glasses to prevent corporate myopia, but some may also need microscopes or telescopes. Differences in viewpoints are appropriate and should be expected. There are no "right answers," only choices. While it is important to understand markets, including customer needs, all firms have to say "no" to some customer requests if they want to stay in business.

The appropriate time horizon for a sales person or design engineer might be hours or days. For a product manager or the manager of a NPD team, it might be weeks or months. For upper management or research managers, it might be years or decades.

In the Information Age, a key to success is productively sharing and merging these disparate viewpoints. The leaders are now "interleaving," planning the next two product generations in parallel, and cross-pollinating information and knowledge between the two.

Back to the Basics #7: Seek and Ye Shall Find

Western management needs to become much more sensitive to behavior patterns, roles, and the limits of power. The first time an upper manager meddles inappropriately in NPD, for example, is when empowerment, local skill, and careful process goes out the window. Even if he can become a hero for solving problems, this is not the best example to set in Information Age firms.

Is it so hard to see that the cost paid for a life of constant crisis is exorbitant? Fixing is the wrong work. It diverts time, energy and attention from that which is important. Today we all know how to best build products of high quality. We all know the folly of attempting to "inspect quality in." If you want quality, it is much better to build it right the first time than to patch it together after it is constructed. Fixing things after the fact is like trying to re-fry eggs, an exercise in futility.

It is astonishing that firms practice these basic quality precepts scrupulously when manufacturing toilet paper, but totally ignore them when leading people or servicing customers. The Machine Age perfected crisis management, but there are more important skills for Information Age firms.

The old way was "management by exception." In that end, doesn't that amount to endlessly fixing problems? One can hardly image a more dismal life or a better way to sap human vitality and physic energy than a constant focus on the negative. Much as dysfunctional, angry people seek out conflicts to feed their hostility, managers seek out problems to solve.

Problem solving and control was the Machine Age career path. The more problems they solved, the more problems managers were given. In some cases managers were paid more to handle the additional work, in some not, but the main point was that problem solving and crisis management was the "important stuff" in business.

Is it any wonder that we have designed systems to generate problems? Machine Age functional bureaucracies ("silos") are superb at generating problems and creating divisiveness. Indeed, all bureaucracies maximize friction and minimize output so that work can increase, providing justification and job security.

We rented a hanger at a small airport, which for 50 years had functioned well with minimum overhead and cost. The city collected rents and hired contractors for repairs and maintenance. Businesses located there, costs were low, life was happy, and all was well with the world.

Then a petty bureaucrat offered to work cheaply to help get the grass mowed and the lights changed. Over a few years, he parlayed this into a full-time well-paid job by generating problems and creating crises. Without his diligent work, those "deadbeats" might avoid payment or break the rules. Rents doubled and conflict was constant as customers were oppressed. When we complained he raised our rent without telling us, then tried to evict us for underpayment. Those who could left, but the airport makes more money than ever. It has a captive market.

The above story has a broader lesson for business: downsize government and distance yourself from parasites if you want to survive. The domestic airline industry lost $9 billion between 1990 and 1993. By 1994 only two firms in the industry were profitable, and one of them was coming out of bankruptcy. The common wisdom is that this was caused by discount pricing, excess capacity, and generally stupid management. The common wisdom is wrong.

The most profitable year the airlines had was 1988. Since then, capacity grew at the modest rate of 1.2% annually. Revenue grew at 3.8%, which was not bad. What grew out of control were government-imposed costs. Landing fees increased at 8.4% CAGR, ticket and passenger taxes grew at 12.6%, and benefits, many mandated by government, grew at 8.4%.

Imposed fees for passenger facilities increased by $1.3 billion, fees for customs and immigration almost doubled (to $775 million a year), and fuel excise taxes of $530 million a year are being imposed in 1995. The airlines are being killed by parasites they cannot escape. It is the same phenomena as our local small airport, but on a Megastate scale. The airlines' profits are being sucked dry by government-owned airports and government-run airway systems.[36]

Business, of course, has its own bureaucrats. In larger organizations, functional bureaucracies are best at generating problems. Each function generates volumes of impressive data to "prove" its good performance. The game functional bureaucrats play is the same as their brothers in government. They protect turf, blamelessly follow procedure, and find or generate problems that can be pushed over their responsibility horizon.

The bureaucrats on the other side of the horizon, of course, can do the same. Thus everyone "fixes" their problems while workloads and staffing levels grow without limit. Government bureaucracy is, of course, the ultimate example, but many companies suffer from the same disease. Some firms even blame their dumb customers for creating problems, at least until the customers find ways to take their business elsewhere.

We accept such behavior because it is the norm, but if we stood back and thought about it we would know that this is craziness. Management gurus have long suggested a focus on ***opportunity***, but few seem to heed. There are clear patterns. In sports or war we try to finish and win, in business we solve problems, and in government we expand bureaucracy by creating and preserving problems.

There are exceptions, but is this behavior not often the case? Today most would agree that government is broken, but is this not also true with too much of what business does?

Incidentally, competitors are not problems. They can be your best friends. They show you where to focus, they give you reasons to change, and they help you innovate. Without competitors, there would be little progress. Without competitors computers would still cost tens of millions of dollars. Without competitors we would still be driving huge, inefficient cars with ugly tail fins.

Viewpoints and behavior patterns matter. Don't people tend to find what they seek and become good at what they practice? If you live by the sword, you will most likely die by the sword. If you live by problems, conflict, and crisis, is it so surprising that you and your firm will likely die that way, sooner or later?

Conversely, try to imagine a world where business runs smoothly, leaders seek opportunity, teams work together, and results are achieved without control, fear, or crisis. That world can exist, but it won't ever be perfect and it won't ever be easy. First leaders must believe, and then they must be able to make vision, empowerment, and self-managing teams work.

We do not say that the goal should be to eliminate crisis, or that crisis management skills are unimportant. Business needs crisis managers, and an occasional crisis serves to stimulate. We merely suggest that if a better balance between crisis management and leadership existed, business performance would increase greatly.

Box 8-11

> **"If you can't do it with brains, you can't do it with hours."**
>
> Clarence "Kelly" Johnson
> Head of Lockheed's famous "Skunk Works"

There is an almost forgotten human side to business: it involves words like "fun," "we," and "winning." Machine Age methods have all too often reduced business to fear, grim process, and a focus on the negative. That is a mistake and a very bad choice.

For a delightfully humorous perspective, rent the videotape, "Joe versus the Volcano." The lesson is that it is better to have fun, make money, and jump into a volcano than to endure the oppressive tedium of the Machine Age. Business should be about having fun and making money. If your work can't be both fun and rewarding, you are doing something wrong, working for the wrong company, or both.

Chapter 9

The Leadership Challenge — Where to Aim?

Key Themes at a Glance

The classical depiction of strategists refers to managerial professionals, totally dedicated and preoccupied with their companies, impersonal in their judgment, and savvy in their business dealings. Similarly, managers control actions and costs. These were the Machine Age roles and models, and they have led us into a world of "denominator management."

Today's Information Age world calls for a different type of leader — one who demands an unwavering commitment to change, learning, and unlearning. The new leader is also caring, smart, and he or she has vision and a clear sense of purpose.

Being a good strategist is not enough. Being a tough manager is not enough. In an age where the Engines of Prosperity drive wealth creation, the leader is faced with the dual task of shaping strategy and providing the energy to make it happen, as well as shaping a social context where change and employee empowerment are paramount.

If there is one thing you should get from this book, it is to discard "coyote" type-training, in favor of that of the roadrunner.

The strengths are different, and so are the mental processes and motivations.

Central to New Think leadership is proficiency in learning and unlearning. Knowledge — technology — is the only sustainable long-term competitive advantage. For over a century, the world's richest person (John D. Rockefeller, the Sultan of Brunei, etc.) has owned oil. Today, for the first time in history, the world's wealthiest person (Bill Gates) owns knowledge.

Nations with large, growing trade surpluses (Japan, Korea, Singapore, etc.) achieve them through knowledge. Major companies today (Intel, Netscape, Microsoft, Sun) own nothing of value except knowledge. Timely knowledge depends on rapid organizational learning, and that, in turn, requires unlearning.

Strategy is All in Your Mind — Literally

Managers and MBA students often want to know the "best" business strategy, so we reviewed the most popular methods (Chapter 5). Which do we recommend? We think executives should be familiar with *all* major business strategies, should read military strategy, and should also collect anecdotal stories of success and failure. Though you don't win wars or become a business leader with textbook strategy, background knowledge helps.

The trick is learning to perceive what can't be seen, so you know what knowledge to apply. That takes experience, and to gain experience you must practice. Top executives and management consultants are well practiced in their fields. They understand how to apply the tools of their craft, and have developed experience in the art of business that gives them intuition for when and how to use which tool. They are able to quickly comprehend and synthesize the whole "system" as it applies to their unique situation and opportunity. The proper approach is an endless learning loop of *gaining knowledge, reflecting, connecting, deciding,* and *doing.*

The skill of value is not so much knowledge itself — it quickly becomes obsolete — but experience, which is the path to wisdom.

Experience lets you apply knowledge in context. It is necessary but not sufficient for holistic understanding and intuition. Rapid and holistic understanding allows informed, effective action — and that, coupled with efficient execution, is how you can win in business today.

The core understanding — a new paradigm — is that business responsibility belongs to the people, not the process. Machine Age firms built process to run the company. Information Age firms use technology and process to help unload people from routine and get them timely information so they can effectively use their knowledge, intuition, and training.

If one considers effective organizational learning loops, some of the ways in which Western education is outdated are obvious. Most MBAs graduate with extensive knowledge, but relatively little experience in connecting and doing. They lack tolerance and patience for reflection, and don't spend time on deep learning. Particularly lacking is teamwork and organizational learning. MBAs are quick at deciding, but their methods tend to be pre-programmed and based on financial abstractions, not intimate content knowledge. Business schools teach process, not intuition.[1]

Box 9-1

> **"I think we have tied acquiring knowledge too much to school."**
>
> Arno Penzias
> Nobel Laureate

Contrast this with the intern or apprentice period required for medical practitioners, lawyers, pilots — and, indeed, most major professions, sports, and arts. It is strange that the concept of a "one minute manager" is so widely and casually accepted. Most would avoid using a one minute lawyer, engineer, brain surgeon, or plumber. Most would think the notion of a one minute athlete or concert pianist was laughable.

The amazing thing is that simplistic rote training without integrated practice and application ever worked well, but it was the foundation for training Machine Age bureaucracies and a major advance for its time. When business was routine, change was slow, and decisions could be extensively reviewed, this form of "scientific management" was dominant.

Refined by Frederick Taylor, it came from the 19th Century Prussian military models. Standard procedures minimized training time and limited how much damage could be done by green junior officers. Predictability and obedience was favored over flexibility and creativity. Inexperienced, obedient, eager junior officers were a useful, if expendable, asset. Promotion came from currying favor and maintaining distance from failure.

Box 9-2

> **"It is impossible to imagine a basketball team learning without practice, or a chamber ensemble learning without rehearsal. Yet, that is exactly what we expect to occur in our organizations."**
>
> Peter Senge
> *The Fifth Discipline Handbook*

In effect Machine Age firms can almost run themselves. Redundancy and routine accommodates high growth and survives massive attrition. It is amazing to behold how much damage, mismanagement, and downsizing a Machine Age firm can survive. Many resemble ponderous, wounded beasts that relentlessly repeat past behavior. You can shoot them in the head with an elephant gun or stick spears in their body, and they keep on coming. Machine Age firms are like battleships, powerful and heavily armored, but obsolete for modern warfare. They are now worth more for scrap than for combat.

Machine Age structures and paradigms still work in some cases. Unfortunately, they do not deal with change well — and that leads to obsolete products, thin margins, and endless downsizing if competition is severe. To enjoy high margins today, firms and products need to be different. Such differentiation takes local intelligence, agility, and constant renewal.

Information Age firms are fragile, but quick and smart. They thrive on chaos, but can't survive Machine Age management. Knowledge-based teams are easy to destroy — they need nurturing, shared vision, mutual trust, and a lack of fear to function.

Information Age business is, of course, different by its very nature. A popular song had a refrain, "Because every hand's a winner, and every hand's a loser..." It conveyed the thought that "the game" was endless and ever changing, requiring constant assessment, insight, and understanding. So it is with business today. Winning in today's product cycle only provides table stakes to play the next hand. Since the next hand is different, this is not Newtonian bench science where the same actions always produce the same results.

Rote strategies and control methods don't work in chaotic environments. Doing what worked *last* time may not work *this* time, and, if it does, it may not *next* time. You only get to invent a new technology once, and the next time it is different. Isn't that obvious?

It didn't used to be that way. The leading companies developed organizational men in their image, and kept repeating the past. DEC repeated its mini-computer strategy, IBM repeated mainframes, Wang did word processors, GM did large cars, Douglas did propeller airliners, and Tektronix did analog oscilloscopes.

Like the dinosaurs and the battleships, the firms were never beaten at what they did best. Still, they all lost in the end to knowledge-based firms who innovated with technology, products, and strategies to change the basis of competition. Though never defeated in their core businesses, the Information Age brought them a worse fate: *irrelevance.* You can recover from a defeat, but the only way a battleship or dinosaur or a Machine Age firm can recover from irrelevance is to change into something else. That's hard.

Let us dispel any thoughts the reader may have that the "winners" have better luck with rote strategy and repetition than the "losers." They do not. Even the "hot" firms have dismal records when they try to repeat what worked in the past. In the 90s, perhaps the hottest and most talent rich industry has been software. Despite spending an average of 26% of sales on R&D, very few software companies — perhaps one in a hundred — achieved significant success with their "second" product.

WordPerfect never got beyond its word processor, and was acquired. Ashton Tate never did anything approaching the successes of its early data base, and faded. Borland never moved past Quattro. Apple's high water mark was their Mac operating system. Lotus had to tap an outside firm to get Notes.

IBM had to buy Lotus to even get into that game. It did so by a hostile takeover at extreme expense and at just about the time when the market was shifting to the Internet, not the Notes, paradigm. The assets "owned" by Lotus (talented people) have mostly left the firm. (Even words give problems: If a firm can own assets, and if its main asset is people, can a firm own people? Most societies and legal systems would say not. Exploring that topic could be worth another book in its own right.)

If anything, Microsoft — who leveraged its DOS dominance to get Windows (after ten years of diligent work and investment), and then leveraged Windows to dominate application software — is the exception that proves the rule. Whether Bill Gates can adapt his insight, business models, and play book of OS-based software strategies to distributed computing or to Internet businesses and applications is an open question.

The lesson should be obvious. *Doing what worked last time may help you extend business success — but only for a while.* The challenge for upper management and corporate boards is ensuring that their firms compete for the future as well as the present. Three things are certain:

- Unless firms constantly renew themselves, they will soon drift into the grim cycle of low margin commodity competition. Once in this cycle, escape is difficult.
- Doing what worked last time rarely succeeds in a new business context. Those whom the gods will destroy are first cursed with success. The more successful firms were, the harder it is for them to change. This happened to the railroads, to RCA, Douglas and Wang.

It is happening to IBM, to AT&T, to the defense firms, and it may someday happen to Microsoft. It did not happen to Hewlett Packard or GE or Intel because they adapted, but it was a near thing, and they know it.

- Wall Street doesn't care. Though it calls itself an "owner," it has repeatedly shown complete indifference as to whether firms survive or not.

In the Information Age, the only strategic resource is empowered minds. These minds are applied to create prosperous new futures. The dominant training methods — case studies, and words of wisdom from the past — are, at best, background information for reflection, not templates for mindless action.

Consider the vast difference between the Japanese and the West's approaches to leadership training. In the West, a CEO with content knowledge — e.g., Andy Grove, Bill Gates, or Jack Welch — is a rare exception, especially in the U.S. or Britain. Legal or accounting backgrounds are the norm. In Japan, CEOs and General Managers with content knowledge are the norm.[2] The Japanese promote managers with hands on experience at creating products, we promote people who manage costs, divvy assets, and make deals. In the words of Mr. Kubota, whose firm had great success in making the transition from the Machine Age — "We bake pies. You divide pies. Which way is best?"

Deal making is, at best, a zero sum game. For every winner there is a loser. Baking pies or making microprocessors is a positive sum game. Value increases, wealth is created, and the pie gets larger.

When training and grooming leaders to create wealth, the leading Japanese firms first hire the best engineers from the best schools. One of many reasons they resist *kodaka* — being hollowed out — is that to recruit first rate talent they have to retain their own product development and R&D capability. In contrast, most U.S. companies don't care.

Once technical experts in a Japanese firm become candidates for management, those who show the most promise will then be given assignments in, say, sales. When the candidate masters that role, they move him elsewhere, perhaps to production or accounting. Before someone is given general management responsibility, he will have both theoretical and experiential learning in all the operational roles of business. Only then, after he has seen the parts of business from many aspects, can he gain holistic understanding.[3]

Western firms occasionally mimic this practice, but their efforts are often superficial. Cynics call this sham "getting your ticket punched." Having sycophants of senior managers rotated through carefully selected assignments so they can demonstrate wondrous results is not the intent of Japanese cross-functional training.

Cross-trained general managers offer competitive advantage. It is not enough to know how to manage, you also must know the technical aspects of what you are managing. Otherwise, people can give you bad information and you will not recognize it as such.

Winning today requires cutting quickly through many confusing things to address precisely the problem or opportunity at hand. Without content knowledge, it is hard to separate cause and effect.

The Japanese take our Western MBA courses, of course, and they are good students. They do well at the course work, but it is not their core purpose. Mostly they want to understand how their opponents are being trained in order to learn how they will react in competitive situations. In many situations, knowing what your competitors will do beforehand is more useful than technology or market knowledge.[4]

Winning at knowledge-based competition depends on smart, quick, action that is tailored to the situation and time at hand. To avoid predictability, you need to practice a variety of plays. Since any predictable strategy can be beaten, you must keep learning and adapting. The learning process is endless. The more chaotic and uncertain the situation, the more important it is to apply the widest possible diversity of knowledge.

You need to keep learning and adapting, and at an ever faster rate. To win today, the following should help:

- Know your business, and keep learning about it. Learning is as important as doing, and effectiveness is at least as important as efficiency.
- Keep trying new things. These must be viewed as experiments. All new products are experiments.
- Be judicious about the use of rote strategies and methods, understanding that global competition will inexorably compress profits on anything that is standard and routine.
- Tap knowledge and experience outside your organization. Beware of knowledge gaps. Staff to weaknesses, not strengths.
- Know that it is the whole that matters, not the parts. Holistic situational understanding gives you competitive advantage.
- Avoid predictability with opponents, but be trustworthy to allies and followers.
- Get rid of abusive managers, even if they are stars. You can't have it both ways.

Empowerment and Envelope Supervision Beats Control

It is amusing, but sad, that "empowerment" has become a misused and hollow buzz-word. Everyone says that they empower people, but almost no one really does it. Most firms just overlay the new trappings on the old ways.[5]

Part of the problem is that empowerment is frightening to managers. One of the few things that functional managers instinctively agree on is that anything that erodes their power is a threat. This has been explored in leadership books, but still more work is needed. Management is about control, but leadership is about shared vision and empowerment.

The root word of management is the Greek *manos*, meaning to handle. Leadership is about journeys and voyages of discovery, not about handling. Systems or machines can be managed, but people must be led. Unfortunately, most Machine Age managers never really recognized that, and they are unlikely to do so at this late date.

A more basic problem is that even if managers could decide they needed to empower workers, most don't know how. They are not trained or practiced in making their associates — "employees," in old think — autonomous and self-reliant. They are trained and practiced in making them serve as efficient and obedient cogs in a machine.

Even when managers outsource, their goal is usually cheaper, not better. The core assumption is that all the best people in the world, except perhaps for a few in obscure specialties, already work for the firm. That's nonsense, of course. The reality is that it is nearly impossible to break outside the box from inside the box, and "outsiders" can offer useful insight.

Machine Age firms avoid risk. Which is most associated with management: (1) "Don't screw up, or else." or (2) "You need to make more mistakes, you are not pushing the envelope enough." All managers know how to criticize and rebuke, but what percentage knows how to coach and nurture?

Consider the dismal job that many managers do in child rearing. The combination of control and neglect ("management by exception") is unhealthy. Control freaks who demand obedience and perfection can drive children to rebellion, drop out, and drugs. We personally know — and everyone whom we know knows more — far too many top managers, some of whom are friends, whose obsession with control has helped create tragically dysfunctional families. We know relatively few who practice empowerment, even for their own children.

We well remember an embarrassing afternoon early in our career, sitting in our boss's office and listening to him grind on at length to his teenage son, using words like "not wanting losers in the family" and "letting us down." We wondered if the boy had committed unnatural acts or gruesome mass murder, but it turned out that he had only finished second in a swim meet.

The father went on to become CEO of a Fortune 500 firm, which he assertively controlled and severely damaged before he was replaced. The son grew up, dropped out, and was last known to be hiding out in the hinterlands of New Zealand.

It does not take rocket science to exploit mistakes in a positive manner for learning and empowerment, but few managers try. It is just not something that is much taught and practiced. Consider a simple three step model to teach your employees (or your children) empowerment and responsibility:

1. ***Don't cry over spilled milk.*** When mistake is made and the damage is done, why not use the wasted material for enjoyment or some other purpose? *Show* them how.
2. ***If you broke it, clean it up or make it good.*** The person who makes a mistake should be held responsible for correcting it. *Teach* them how.
3. ***Don't make the same mistake twice.*** The person who makes a mistake should do a post mortum and learn how to avoid it next time. *Ask* them how.[6]

These used to be accepted core principals to live by, but somehow Western managers have lost them in both their personal and business lives. Even Western religions have slowly slipped into the model of a wrathful God who says "Don't screw up, or else." Engineering used to be based on experimentation — making learning mistakes — but lately the emphasis has been on copying and using safe procedures to avoid mistakes. Mistakes in Machine Age firms are associated with blame, not with learning.

Managers stand at a fork in the road. Is life about following rules or making wise choices? The first path, now a well lit and paved superhighway, leads us inexorably to Machine Age organizations. The second path, one less well marked, takes us into the Information Age.

It is core fact, that high effectiveness organizations cannot work with low integrity people. Information Age firms depend on a politically incorrect notion — all people are <u>not</u> equal. Each person has unique talents and abilities. Unfortunately, some talents come from the dark side. Machine Age firms often tolerated negative behavior — deceit, dishonesty, unethical behavior — so long as it got results and was not proven to be illegal.

The headlines are filled with stories of athletes' wrong doings, insider trading scams, corporate greed, and political shenanigans. Since this behavior destroys high performance organizations, it cannot be tolerated. Indeed, U.S. government is so distrusted that the majority of citizens no longer bother to vote. If Constitutional government is "of the people, by the people, and for the people," then what is left when the people, with cause, distrust government? What's left is a looming crisis, and the same holds for business.

Teams by definition depend on mutual trust and the ability to depend on others.[7] The utmost attention should be devoted to the selection of highly qualified people with positive values. This gets us back to the older values — putting "the best person" in the job. It moves us away from the 1960s notion of quotas to ensure that people with minimum qualifications were selected before those with superior training, ethics and talent. Expect social conflict as civil rights clashes with the need for global competitiveness.

What holds for teams is even more true for leaders. The learning organization model of leadership is totally different from the prevailing model of the charismatic, cheerleading, "personality ethic" leader. Leadership is not about "good hair" and cute sound bites, it is about integrity, competency, honesty, and vision. Stephen Covey's book on the seven habits for highly effective people should be considered as must reading for leaders.[8]

- A leader without followers is not a leader. A team without open, honest interaction is not a team.
- A good test for empowerment versus control: Can your teams function, and well, and make effective decisions, in the absence of their leader?

Machine Age managers tend to scoff at empowerment saying, "Who makes the tough decisions?" Implicit in this is the assumption that a team is incapable of allocating painful effects, that it can't handle layoffs or pay cuts, and that it can't demand performance from its members. We disagree. It is exactly because these decisions are so critical and involve so many people that leaders should demand involvement from the people who will be affected or held accountable.

There is much evidence that self-imposed honesty and empowerment is better at tough decisions than control. In the 20s, when investment advisors made major blunders, some felt responsible enough that they leapt from windows. In the 80s, the Captain of a Japanese ship carrying cars to the U.S. made a mistake that caused a hold to flood, severely damaging the cargo. He committed suicide. Today the system churns, the government regulates, laws and lawyers multiply, the courts overload, but, in the end, it usually turns out that no one is held responsible for anything. The old rule-based system drives a feeding frenzy of litigation, but results are poor.

One excellent learning case is "Freddy," a final assembler at a manufacturing plant. Freddy's component was hidden inside of a much larger finished product, but was critical to its performance. If the end product failed it not only caused the loss of some very expensive equipment, but many lives. The equipment was used in nuclear power plants.[9]

Management and the team designed in many safety checks. They took the safety problem seriously, and knew they were responsible for it. The problem was that Freddy was an active alcoholic, and a clever one. He missed enough work to be put on warning, but not enough to be fired. He was suspected to drink on the job, but it couldn't be proven.

One day Freddy crossed the line. He showed up obviously unfit for work, and was intercepted by his teammates and forced to stay in the locker room until he sobered up. The manager was determined to practice leadership, so he called an emergency "work stop" team meeting. Work would not resume until the Freddy problem was solved.

Freddy, of course, started with denial. He started to storm out of the room, but was blocked by the team. "Sit down and shut up, Freddy. We've got a problem and you are going to stay and help us fix it. We're not going to let you kill yourself and us, and who knows who else."

They formed a "Save Freddy" brigade, and started discussing options, which centered around helping Freddy stop drinking. Freddy was the brigade President. The team's leader provided information on recovery programs and meeting times. Some members arranged a schedule where Freddy could go to two AA meetings per day. Others donated personal time to cover the weekends. All agreed that before anything could start, Freddy had to admit to his problem and "agree to do whatever was necessary to stop drinking."

After a long, embarrassing silence, with all eyes on Freddy, he admitted his problem and agreed to do whatever was necessary to stop drinking. He did, the team made sure he did, and that story has a happy ending. Freddy did stop drinking, one day at a time, and he became a good worker and eventually a general foreman.

This example was deemed to hold four lessons for leaders. First, of course, is the need to set limits, to draw clear lines in the sand. The Freddy problem could not be allowed to continue. Second is the need for tough standards, in this case reminding everyone on the team that they were responsible for lives. Third is the leader's responsibility for providing information and options. Fourth is the leader's responsibility to set an example by his personal, active involvement. The leader accompanied Freddy to his first AA meeting and volunteered his time for Sunday of the first week.

The example concludes: "Leaders don't just lob the ball into the person's (or the team's) court and then just walk away. That is not leadership. That's abdication. Leaders continue to coach and support because they are genuinely interested in the individual's success."

Box 9-3

> **"Leadership appears to be the art of getting others to *want to* do something that you are convinced should be done."**
>
> Vance Packard
> *The Pyramid Climbers*

The Freddy example, of course, could have been handled with rules, control — and probably proof and litigation after disaster strikes. The

Exxon case comes to mind. That is the traditional way, but is it better than leadership and empowerment?

We think not. Proof, control and micro-management are good ways to guarantee that your business will decline. Many firms prove that annually. They struggle, they cut, they intimidate, they chase fads, but their products are consistently mediocre, or worse, and late.

Less can be more. Consider the U.S. airline industry. Cut to the bone, it limps along awash in red ink. Planes age, maintenance is deferred, mechanics are laid off, and employees are squeezed to the limit. This is lean and mean at its extreme. On top of this, in the name of safety, aircraft operations are regulated to the hilt by government bureaucrats. You can't change a bolt or start an engine without the right paperwork.

In theory, the bureaucracy ensures safety. In practice it moves service ever downward, and business travel get worse every year. Some say the remaining differences between low fare airlines and Mexican buses are because the U.S. animal rights activists would never allow chickens and pigs to suffer such abuse.

Employees live in fear. Flunk a drug test because you ate a hamburger with a poppy seed bun, and you might be grounded. Break a rule and your career might be over, but start engines and "block" your 747 in eight minutes.

In contrast, consider Japan Air Lines (JAL). They are subject to the same economics and regulations, but they don't react the same way. Procedures are necessary, but trust is sacred and foremost.

JAL lost a 747 a few years ago in a tragic accident caused by a mechanical failure. 520 people died, but how the firm responded is a good example of trust and total team commitment.

The President of JAL got into his limo and personally visited relatives of the victims, hundreds of families. He sincerely apologized to each of them for the accident. Then he resigned. Because he was responsible.

JAL dedicated some 500 employees to helping grieving relatives. They took the families of those who had died to Mount Osutaka where the plane crashed. In some cases they literally carried them up the mountain to visit the crash spot. They did this for years on each weekend and on the anniversary dates of the crash.

It was clear that a solemn trust had been violated and that JAL and its customers had suffered serious damage. The error was maintenance related. Several with that responsibility for that specific plane committed suicide, some years later and after having cared for the bereaved families. As with the JAL President, simply knowing they had screwed up and let the team down drove all the corrective action and atonement necessary.

Which system would *you* prefer: (a) total team trust and commitment, or (b) abuse, procedures, and bureaucracy? If you had a choice, whose aircraft would you fly? Whose products would you buy? Who would you trust to run your local nuclear plant?

Contrast JAL's action to the traditional U.S. combination of micro-management and bureaucracy. The Hanford reservation in Washington State dumped radioactive materials into the environment for several decades. This broke no rules at the time.

Victims in surrounding towns now have the impossible task of "proving" that their specific cancers stemmed from radiation releases long ago. This is the "tobacco defense," and our system often breaks down into such acrimonious and unproductive conflict between hair-splitting lawyers and expert witnesses. Trust is low, so litigation and gridlock results. Billions are wasted that way. In all the law and bureaucracy, justice sometimes gets lost.

Small is Good—From Markets to Niches

We are trained to think big, and miss the profitable small opportunities as a result. In the 60s, the cold war was at its height and big was all that mattered. If there were social problems, then the solution was big government programs. If there were business problems, the solution was for big business to apply massive resource. The old General Motors business model still worked. All markets were mass markets, and the approach was to saturate them with volume production and mass merchandising of adequate products. "What was good for General Motors was good for the country," and, "No one ever got fired for buying IBM" were the slogans of the day.

As recently as 1970, U.S. leaders were still seeing "a light at the end of the tunnel in Vietnam." As recently as 1975, Detroit dismissed the threat of small foreign cars. As late as 1980, bureaucrats still spoke of "winning the war on poverty" through massive social programs. In 1985 the New York Times ran an article under the headline "The Daunting Power of IBM: It keeps growing stronger."

In 1990, Soviet Russia was still a threat and NASA arguably was still used by some as a model for technology development. In 1992, Clinton was elected because he claimed government could "fix" the economy, a promise he has had problems keeping. Even in 1995, officials who question the effectiveness of the government's colossal ($250 billion spent so far!) "war on drugs" shouted into silence. The public, lacking solutions and resigned to the Megastate doing as it chooses, for the most part doesn't waste its time on "defense conversion" or "drug war" debates.

These changes, when viewed as a time lapse photograph, are obvious beyond doubt. The dominant machines of the Machine Age — large cars with tail fins, massive computers with spinning tape drives, and, yes, even the space shuttle — today have much the same quaintness that one would experience when examining a museum of 19th century locomotives. The planning of the era was as quaint. Machine Age firms demanded "provable" $100 million plus markets in two years before they would invest. As a result, they almost always missed the emerging opportunities.

The same environmental changes have wreaked havoc on social organizations. Thanks to "installed base" (i.e., citizens made dependent upon government social services), tax revenues, and brute force, Megastate bureaucracy persists, but the trends are clear. Here too is an opportunity for the private sector. Imagine what competition could do for education if only control could be wrested from the government.

The fragmented bureaucracy of the U.S. won out over the centralized control of Russia, but it was in turn outperformed by governments favoring smallness, indirect influence, and timely knowledge. It is often said that the smartest people in Japan work for MITI, but few would say that about the U.S. Department of Commerce.

Box 9-4

"It probably will not come as a surprise that less innovating companies are dominated by tall hierarchies, and that honoring the chain of command is a value."

Rosabeth Moss Kanter
The Change Masters

Many books can and will be written about the power shift from the big to the small. Clearly, today the small dominates. Nowhere is this more true than when discussing market trends.[10]

Why is this true for markets? Why have all markets changed from generic mass markets to specialized niches? We submit the major differences between the ponderous largeness of Machine Age industrial leaders and the small quick predators of the Information Age are the result of three factors — time, knowledge, and technology.

It starts with the fact that many customers are well informed. Thanks to global communications and computer technology, small firms and even consumers are often more in touch with trends than large organizations.

There is a new market segment in Personal Computers, the "heat seekers." These are smart professionals who have started small businesses. Heat seekers almost never visit computer stores, because they know more than the clerks and don't want to pay retail prices. They are on top of the technology and they want leading edge performance, but they are also cost constrained and price sensitive.

They use their knowledge to specify the hottest systems, right down to memory chips and internal bus structures. Then they have their computers solicit twenty or a hundred quotes for the specified system. They can buy from anywhere in the world, electronically.

So customers themselves are using time, knowledge, and technology to get exactly what they want, and when they want it. This is happening in all markets, from food staples, to entertainment systems, to industrial equipment. Shopping and purchase practices have changed.

For example, when the authors needed Christmas presents, we purchased exotic CDs through a national computer database conveniently and at low cost. We used specialty catalogs. We bought Oregon craft products through a local mass merchandising warehouse, and then we had these gifts shipped by a local private post office. We avoided malls, saving money and time. We buy books off the Internet with the click of a mouse, enjoying wider selections and lower prices.

Large firms are copying these practices. Intel lets project teams purchase needed items on credit cards, because it is faster and better than using a ponderous, generic, low content-knowledge, central purchasing organization.

On the other side of the equation, it takes massive firms to serve mass markets, and that attempt is becoming a losing proposition. Once, size was an advantage, and there was major impact from economies of scale. A large firm could purchase parts or distribute products much cheaper than a small company. It had access to internal talent and knowledge beyond the wildest dreams of its competition.

Today these advantages are less clear. Small firms can get very attractive prices on components, and even if their larger adversaries can do a bit better, it matters less. Component costs are now small fractions of the cost of most manufactured products in the dynamo markets. In electronics, the parts cost of most products is in the 10–15% range.

As for timely knowledge, smallness can help. There is a truism, now generally accepted, called Joy's law, after Bill Joy, Sun Microsystems' Vice President of technology, "No matter what's the name of your company, most of the best talent in the world does not work for you."

Consultants, contract workers, and design firms are available that can often get the job done better and cheaper than internal talent. More than half of North American firms now use consultants on a regular basis to help with their product development. In similar manner, the trend is toward "virtual companies" who contract for manufacturing capacity, distribution, or whatever. Virtual corporations are still hazy models and there are major problems in locating trustworthy partners under the West's law based system, but the future is clear.

Often technology allows exceptional performance at inexpensive prices because vendors agree to support standards. Customers no longer have to depend on one mega-firm to do everything well. Instead, using "open architecture" allows customers to get the best of everything from a variety of suppliers, and still be assured that all the parts work well together.

Today's model is personal computers. You can buy the best display from one firm, the best graphics board from another, and the best hard disk or keyboard from a third. Because of standards and intense competition, the end user can be fairly certain that the parts will work as a total system. If he wishes, he can buy such a custom configured system from a local computer store, and they will put it together for him and provide product support. For complex systems, there are Value Added Resellers (VARs) that will configure large, networked, computer systems, make them work, and even provide user training.

Tomorrow's model is even more open — the Internet. If there was ever a model of standards setting under a level of chaos approaching anarchy, this is it. Unless the government starts taxing or regulating the Internet, this is the next level of business evolution. Not only are the firms virtual, but the market place itself is becoming virtual.

There has never been a better time for the customer. Why settle for mass-market, generic, one-size-fits-all products, when you can get exactly what you want at an affordable price? So all markets are niches, and experts see niches within niches, many levels deep. Within the PC market, which is niched by performance level and platform type, there are niches for add-ins, and niches for software, and niches for software add-ins.

In this one market area, personal computers, the Comdex trade show (recently purchased by a Japanese firm) draws almost

200,000 savvy customers and fills all of Las Vegas with thousands of firms displaying new products. The knowledgeable user can purchase exactly what he wants. The neophyte can easily get a system that works.

It almost does not matter any longer to the customer what brand name is finally affixed to the system. It matters greatly to the vendors, and they compete ferociously for differential advantage. Vendors niche frantically to gain recognition and raise margins.

Dominance to "Nimble-and-Quick"

When the history books of the next century are written, a date will be set for the transition from the industrial age to the information age, the transition from a Capitalist to a Post-Capitalist Society. The exact date is debatable. Peter Drucker uses 1989-90, *Fortune* says 1991, and we prefer October of 1981. In any case, it has already occurred.

Industrial age competition was dominated by size, power, and what IBM called the FUD factor. FUD stands for Fear, Uncertainty, and Doubt. To a large extent, FUD is what guided product purchases, especially in high technology areas.

In the industrial age, each market had only a small number of competitors. Each country had things that they specialized in, and buyers were conservative. If you wanted a computer for business, you bought from IBM. If you wanted electronic test equipment, you bought from Tektronix or Hewlett Packard.

If you wanted a good watch, you picked a Swiss brand. If you wanted consumer electronics, you bought RCA, Zenith, or other U.S. brands. If you wanted a good wine, you bought a French label.

The average American consumer bought a U.S. brand of automobile, probably the one their father drove. If you were an eccentric, you bought a British car and worked on it each weekend. If you wanted prestige, you could buy Mercedes or Rolls Royce. If you wanted to make a social statement, you bought a Volkswagen and painted daisies on it.

There must have been some products that the Japanese made in those days. We are hard pressed to remember much but Sony radios and cheap toys. There wasn't much product differentiation, and what there was seemed to be based on technical specifications or fashion. It seems silly now, but customers of that era bought things like "six transistor radios." If one did switch to a different brand of automobile, it was probably because of a horsepower or styling advantage.

What is amazing in retrospect was that if a firm fell behind in a product competition, it made little difference. Customers doggedly stuck with the brand. They trusted the dominant supplier. When necessary, they delayed purchases and waited for the leader to catch up.

In the 60s, Tektronix was the world leader in oscilloscopes, the major type of electronic test equipment. Their products were large and hand crafted. They were built from vacuum tubes and esoteric in-house components.

Hewlett Packard, for strategic reasons, needed a presence in oscilloscopes. They developed a more modern product line. Their products were smaller and better, and were built from Information Age transistors, not Machine Age vacuum tubes.

Everyone knew that HP products were better. Tektronix knew, HP knew, and the customers knew. But they also knew that Tek would eventually respond to the challenge. So customers waited patiently for years — several years — until Tek introduced competitive oscilloscopes. There was little or no change in market share.

The same was true in computers for many years. IBM never led in innovation or price-performance, but they were the safe choice.[11] Company after company assaulted IBM's fortress, only to have their shattered bodies thrown back from the walls.

Competition was almost a ritual, with the outcome preordained. IBM's main competitors of the era were known as the BUNCH (Burroughs, Univac, NCR, Control Data, and Honeywell). None were serious threats, and neither was GE or RCA. One simply didn't switch brands.

Doing business in the 90s is nothing like this. Today there are many competent suppliers, and an overwhelming selection of products. Life cycles for products are so short they are not measured in years, but in weeks or months. New products, if done properly, are so compelling that customers feel the strongest pressures to upgrade.

Intel's 386 was the dominant product in 1991, but it could barely run 1993 application programs. Most run under the Windows interface, and Windows running on a 386 gives one an entirely new definition of "patience."[12]

It was inexpensive to upgrade to a 486 system, so most users did. They did it in business to be competitive, and they did it at home to run the latest games with interactive sound and video. By 1994 heavy users were moving to the Pentium, and by 1995 it was the standard choice and vendors who were not aggressive at shifting (e.g., Compaq and IBM) lost share. The notion of Windows '95 running on a 386 was almost a joke.

Books are being written on time-based competition.[13] If a vendor misses a new product cycle, they are often locked out of the market, no matter what their reputation.

McKinsey & Co. was the first to publish the obvious (1983) saying "Six months of delay can reduce a product's life cycle profits by 33%."[14] They ran a spreadsheet to show that, for most markets, nothing correlates to profits more than time to market. Compelling new products allow high margins, but prices erode rapidly and life cycles are short.

Today it is common wisdom, and true, that firms market early with a compelling new product prosper, while latecomers usually lose their shirts. Note we said "early," not necessarily "first." Companies blunder if they release products that are unfinished or disappoint customers. This is becoming a common and costly mistake.

Hewlett Packard, a world-class global competitor, carefully tracked the time it took for their new products to fall to half their peak revenue. The average time to half-revenue decreased by 50% from 1979 to 1988.[15]

A mistargeted or late-to-market product rarely pays back its development cost, and failure can destroy careers and companies.

Machine Age firms, especially those that have been victims of "slash and burn" layoffs, avoid risk and innovation. This has led to the "clone wars," where all products are imitations, margins are slim, and downsizing is continual.

Box 9-5

> **"But the true 'hollowing' of America is the loss of technological and innovative leadership, supposedly America's long-term competitive advantage, because of a stubborn refusal to face up to the core of the problem — long new product development and introduction cycles."**
>
> Stalk and Hout
> *Competing Against Time*

Machine Age firms and models just can't cope with fast cycle competition. No firm in the world was better at industrial age competition than IBM, but the very things that once made them successful work against them in the new era.

IBM was good at preventing mistakes and driving new trends. They looked only for emerging markets big enough to fit their techniques and strengths. They wanted markets that could be clearly identified and sized in business terms.

Unfortunately, that almost prevented IBM's being early with new technologies, and it blocked doing new products that might threaten their existing large product lines. An even greater trap was their successful past. It caused arrogance and reluctance to change.[16]

In the late '90s, IBM is still formidable, but fading and time is running out. It is strong in the old markets, but weak in the new. Its future depends on leaving its fortresses and attacking Information Age firms on their home turf. Its Army still outnumbers all the rest, but how and where should it attack? To have much of a future, it must beat Microsoft in PC operating systems, or Oracle in networked databases, or Novelle in networking PCs.

These markets didn't even exist in the golden age of IBM. It has not won major battles in any of these, and many IBM troops are committed to holding actions in mainframes and PC platforms. When he took the job, Gerstner, IBM's CEO, said IBM needed "basic blocking and tackling, not vision." We think he is wrong about that.

How to Explain What to Do?

For over a decade we have been seeking a simple model to show business leaders what to do for success. The type of learning and skills that are needed don't even fit into Machine Age models. Sending managers to another TQM or financial management course is not the answer to winning in the machine age. What's needed is a different viewpoint, not more te chnique.

Recently we were reminded of a simple, well known metaphor that may help.[17] Remember the childhood Roadrunner cartoons, which pitted the strength and planning of Wile E. Coyote against the speed and agility of the Roadrunner?

The powerful Coyote always lost but kept slugging away. He is almost a symbol of the old Machine Age firms and their methods. He had a sole-source vendor (Acme), was goal focused, and a relentless competitor. His life was devoted to catching the Roadrunner.

He was not stupid, and he was not weak. Each time he developed a detailed plan and followed it to the letter. Each time his "catch the bird" contraption was accurately and methodically assembled and placed, but it never worked. Children want to scream at the coyote to stop following procedure and get a life, but he never did.

In contrast, the Roadrunner was intuitive and ever so quick. He did not just escape, he was never threatened by all the power, mechanical devices, and explosives that coyote could wield. He was never seriously concerned. He sailed through trap after trap, and disappeared over the horizon with a joyous "beep beep."

Coyotes follow procedure. They have competencies and skills but are caught in an endless cycle of "develop plan, set trap, get tricked, get bruised." Roadrunners adapt, experiment, and change the rules in real time. Roadrunners are magic: poetry in motion at warp speed.

The coyote is the Machine Age manager: stolid, resourceful, methodical, solemn, diligent, and ever so persistent. The roadrunner is the entrepreneur, the creator: joyous in his freedom and agility of movement. He sees realities that coyote can't imagine. When coyote paints a tunnel on solid rock, the roadrunner, with a leap of faith, dashes through the tunnel. When coyote, astonished, tries to follow, he smashes his face.

If you get only one thing from this book, let it be that we don't suggest giving more coyote training for machine age managers. We suggest long term coaching and roadrunner apprenticeship programs. You don't make a coyote into a roadrunner by sending him to a three day course, and not every coyote should aspire to becoming a roadrunner.

The basic approach to business needs to change. Machine Age firms needed earnest, stolid managers. Information age firms need passionate, flexible leaders. The basic ingredients are joy, exuberance, compassion, and sincerity. If you fake these things, it will not work.

The strengths are different. Coyotes are persistent and resilient. Roadrunners are resourceful, adaptive, and, above all, quick. They don't follow plans and procedures mindlessly, they experiment continually, adapt creatively, and live uniquely. They don't accept coyotes' limited view of the world.

The mental processes are different. Coyotes are smart, but they work analytically and methodically, using linear logic. Roadrunners are embodiments of the ancient Greek concepts of wisdom through total awareness. They don't analyze, they intuit and act quickly to exploit emerging reality.

The motivation is different, and this may be the hardest thing to change in Machine Age firms. Coyotes operate from what they want, "I want to catch the bird." In Machine Age think, "I need a profit margin of 23.7 percent and a growth rate of 14.3 percent." The coyote is obsessed with beating the competition, with winning. His entire focus is on external reward. He is the ultimate competitor, and the ultimate loser.

The roadrunner operates from who he is. He barely notices coyote's pathetic traps. He stands for something, he does something, and he does it world-class. In the cartoon, the roadrunner is a magical creature of speed, mobility, and maneuverability. He is the best in the world at what he does, and his whole demeanor shows that he knows it. Is this different from Nike in shoes, Wal-Mart in retailing, or Intel in microprocessors for the desktop? How many firms in their heyday exuded such exuberance and confidence? What happened to them when they lost that feeling?

The real irony, of course, is that a confrontation between Coyote and Roadrunner in a closed building would have a different outcome. Roadrunners are defenseless from coyotes in their own firms.

Long ago, one of the authors had a conversation with Howard Vollum, a man of genius and compassion, a man who became wealthy by being the co-founder of Tektronix, which was, in his life, well respected and unexcelled at what it did. We asked what he thought of the then popular business strategy of managing by market share, as popularized by Boston Consulting Group. He was retired, and Tektronix had fully embraced BCG's textbook strategy and most other precepts of Machine Age management.[18]

Howard was quiet for a long time, and he looked vaguely sad. Finally he said in a very soft voice, "I think they got it backwards. If your products are excellent for the customers, and if your firm is excellent at how it treats the customers, then you get market share and profit. It is the excellence that makes you prosperous, not the market share."

This was wisdom. Howard was a roadrunner, but his now large firm was being managed by coyotes. The last time we spoke with Howard was in the early 80s when he served on the board of a new venture.[19] His health was failing, but he came to almost every meeting, where he would revel in the exciting new products and tales of delighted customers. He said, "That's why I like to invest in start-ups. This is fun. If I worried about Tektronix, I'd have another heart attack."

Howard died a few years later. Today his firm has downsized to about a quarter of the employees it had at its peak. It's about the same size it was in the 1980s when inflation is considered an average in its industry. That's typical, and about the best coyotes can do.

When we speak of "micro strategy," we mean options for personal action. There are many things that you can do for yourself to make your life and your career better. The key understanding is that choices exist for individuals, groups, companies, and countries. The world, including the world of business, is not a massive machine designed by a malevolent God that grinds along despite our intellect or our willpower. We are all co-creators of our future, and it starts with imagining how we choose it to be.

The slogan carved over the gates of the Nazi concentration camps, "Work shall make you free," was a lie. There is little evidence that mindless work is rewarded or productive. There are choices. Where you stand today is to a large extent determined by how you have chosen to live, and this will be ever more true in the future. If you focus on fear, toil, control, and scarcity, that is, most likely, what you will get.

We keep on our desk a crystal star engraved with the words, "We create our own future." Somehow most people keep missing this point. There are multiple realities, and we get to create them. There is no one winning product, there are thousands, perhaps millions. There is not one future, but hundreds, perhaps thousands. There is no one career path, there are many. Each day we get to make choices and decisions that shape our futures.

One of the questions that we were asked repeatedly by managers who reviewed drafts of this book was, "Are you crazy? I work for a traditional firm. We are doing OK, because my job is to make sure we do OK. Why do you say that traditional business paradigms, and the people and firms that embrace them, will lose, when today they are so totally dominant?"

We don't say the Machine Age firms will all lose or always lose. We merely note that few are happy places, and most have not been doing well in the markets that matter. We, and many others, say the old ways are less relevant in high prosperity technology-market areas.

Some of old line Machine Age corporations have managed to renew themselves. Hewlett Packard and Motorola are excellent examples. They are not perfect, but they have prospered by becoming very different organisms that they were a decade or even a few years ago.

We don't suggest that small innovators attempt direct toe-to-toe confrontations with Machine Age Megafirms or Megastates. Mammals proved superior to dinosaurs, but not by being stupid. Mammals that made direct assault on a Tyrannosaurus Rex would not live to contribute to the gene pool.

Still, before the mammals came downsizing. Smaller, nastier saurians like the velociraptor evolved first. They were better adapted and could easily steal food from their larger brothers. It took many such cycles before an ecological niche for mammals with brains developed. Today's equivalent of velociraptor packs savage Machine Age firms via hostile takeovers, stripping them to the bone of cash and then selling the assets. Takeovers hit $350 billion in the late 80s, and a new record was expected in 1995.[20] Still, these periods of carnage exemplify bloody transitions, not models of viable futures.

Paradigm shifts are times of discontinuity, times when predicting from the past fails. Evolution is marked by episodes of revolutionary change, and this is one. From the dinosaurs, to Rome, to Spain, to Britain, to General Motors, Firestone, Pan Am, and IBM, history has proven repeatedly that past success is a poor predictor for the future. In such times past success is more a curse to be overcome than a blessing — it makes you a target. The Information Age transition is already well established by every observable metric. The Fortune 500 has lost jobs every year since 1970.

Still, the change to Information Age methods won't happen because we say so. In the end, *you* get to decide. *You* have choices. *You* can think about what we say, assess the situation, and decide where to place your bets and how you want to live. *You* can choose to work in Megastate bureaucracy, you can choose a firm locked into Machine Age paradigms, or you can choose one with an Information Age view of the future.

The organisms and paradigms that offer the most survival value will prevail in the end, but it takes time, often a long time, for a complete ecological change. Business is a rich ecology, and the Machine age paradigms still function. There will always be markets for commodity products.

Everyone loves buying cheap clones, and there is always someone willing to work a little quicker or cheaper than the dominant competitor. Like the dinosaurs, Machine Age firms are being more savaged by their progeny than by Information Age competitors. IBM, for example, is losing the PC market to clones, not to firms with new concepts of computing. There are many upper managers of successful new ventures who use Machine Age fear, process, and control. They chortle over how the older, larger firms can't compete because of higher paid employees and benefits.

Traditional firms can be good (cheap, fast) suppliers and good (needy) customers for knowledge-based companies. Machine Age and Information age firms need not be competitors, in fact, they can become symbiotic partners. The age of the dinosaurs is eons past, but there are still many species in existence. They are just less relevant to the mainstream. So it will be with firms who use traditional Machine Age methods.

Once you accept the fact you have meaningful choices, you need to sort out what to do. If you are going to work smart, you have to know what you can depend on and what scenarios are likely. If you think deeply, many of the important things are predictable even in an age of chaos.

What assumptions might be safe bets, and lead to competitive advantage? How might these be extended? Here are a few simple examples of useful scenario building. We suggest this is a skill to practice, both at the individual and corporate level.

- Bet on Moore's law. The density of micro chips will double every 18 months. This gives the same performance at half the price, or twice the performance at the same price.

  Extension: The same sort of non-linear trend should hold for the cost of bandwidth, as soon as communications are moved to the free market. Where might this happen first? In unregulated areas, such as LANs and private networks.

Implication #1, there will be a myriad of profitable niches, such as providing the high bandwidth services of corporate America to small business. Implication #2, From newspapers to record stores, old line businesses will be threatened by this trend. Who wants to drive to a store to rent a videotape, when you can instantly download exactly what you want, and for less money?

- Bet on the reluctance of managers to change. Assume an infinite supply of firms who will supply cheap commodity products and services. These will bid with thin margins, and will drop prices below costs to meet competition.

  Extension #1: Knowledge and uniqueness will become ever more valuable for competitive advantage.
  Implication #1, knowledge thefts and industrial espionage will increase (it already is).
  Implication #2, there will be lucrative opportunities for knowledge workers worthy of trust, but as both supply and demand increase it will be harder to know who to trust.
  Implication #3, Machine Age firms will seek ever cheaper workers, having them work in ignorance.
  Implication #4, certification and codes of ethics will become more important for knowledge workers.
  Extension #2: Mergers and acquisitions will continue unabated.
  Implication #1, as firms internally lack the resource and diverse experience for due diligence, this will move to specialists.
  Implication #2, that will cause more litigation. Interesting question: Will litigation start targeting law and accounting firms? Probably.
  Implication #3, there will be more regulation of mergers, stock trading, etc.

- Bet on continued downsizing from firms that have chosen that option. Layoffs are addictive. If you had one this year, you are very likely (63%) to have one next year, regardless of the results.[21] Once a firm is pushed back from the cutting edge of market and technology knowledge, it lacks the margins to invest much in regaining this knowledge.

> Extension #1: High value added in Machine Age firms will mostly be outsourced. By the year 2000, most workers will be outside the firm.
> Implication #1, Outside firms, mostly small, will provide crucial value added — like innovation, new product development, and MIS services — to Machine Age corporations.
> Implication #2, these services will be value priced, but the cost will be higher than many traditional firms will want to pay. This will tend to drive more contingency or "piece of the action" contracts.
> Implication #3, This will expose major gaps in Western tax and property law for knowledge assets.
>
> Extension #2: The gap between the haves (Group A) and have nots (Group B) will become an ever widening chasm. It will grow into a Grand Canyon, for workers, companies, and nations. This trend will drive a number of self-reinforcing "feedback loops."
> Implication #1, brain companies (Group A) will invest heavily in growth, training, and new technology, creating an upward spiral.
> Implication #2, brawn companies (Group B) will be pressured to return higher dividends to stockholders, leaving less for training and technology, creating a downward spiral.
> Implication #3, the cultural gap between citizens of Group A and B will grow. Since wealthy knowledge workers and illiterate, angry barbarians will be voting citizens and near-neighbors, the stress on Western democracies will increase. (Already U.S. cities have become more dangerous than the wilderness, the first time in history this has ever happened.)

We could continue ad infinitum, but you get the idea. You start with the near-certain assumptions that overlap your business, and then you place them in context and use experience and intuition to see where they lead. Rather than building plans to allow control, you develop scenarios to allow insight (and outsight). For best results, you should tap external sources (remember outsight), but you can start this thought process internally.

The next step is left as an exercise for you. If the above trends interest you, practice refinement and sorting through scenarios to find business opportunity. What can you find that might increase your firm's sales and profits? If you disagree with our listing or find it irrelevant, create your own and use it to explore possible futures.

Here are some actual application results, based on the above scenarios. We teach, write, and started a consulting firm to help selected clients with Information Age business issues. A friend started a very successful outplacement business.

Many businesses target people's needs for safety, and these are not all firms selling security services or weapons. People are starting to "cocoon." The explosion of service businesses like take out food and videotape rental stem from this trend. Doesn't this also drive markets for video-on-demand and large screen TV sets? A friend provides armored limousines to foreign executives who consider running U.S. assembly plants as hazardous duty.

Of course, not all aspects of the future are obvious. Some of the best business opportunities are gambles. You can bet that executives would love to transact more of business electronically, if that option was universally available, affordable, and secure. Business travel is increasingly expensive, inconvenient, wasteful of time, tiring, and even dangerous.

You can also bet that strong forces will oppose and try to usurp this trend. The Megastate will delay what Andy Grove calls "rich media" if it can — and it can. It regulates the infrastructure needed for high bandwidth communication. It would like to own it outright, as it does with Machine Age infrastructure. Many states run phone and cable companies, using public funds to compete with private firms. Many U.S. government agencies oppose electronic privacy rights.

Should it be a surprise that bureaucrats are blocking Information Age infrastructure, overtly or covertly, at every level, and with every fiber of their beings? They have delayed the deployment of inexpensive local fiber and digital services for decades. Except for the large firms who can afford private networks, the U.S. lags the world in this crucial transition.

Governments are terrified because the tax revenue that fuels the bureaucracy is at risk. Few in government understand the new technology and its potential to create a golden age of prosperity, but many see how it can threaten them. Physical commerce is easy to tax, police, and control. Ports, airports, roads, buildings, and all forms of physical infrastructure are easily controlled or taxed by the Megastate. Information Age assets are more elusive.

We FAX instead of mailing. We exchange data files and download software. We order goods and provide customer service electronically. We are starting to video conference. Trillions of dollars dodge government toll gates by electronic fund transfers. The Megastate wants to shift its taxes and fees to electronic commerce, but a shifting of billions of dollars in taxes will be noticed and opposed. This scenario is still evolving, and will be interesting to watch.

Scenario building was not a skill much used in the Machine Age, but it is one that will be honed and perfected in the Information Age. Already a few groups in firms like AT&T are collecting knowledge bases of experience for their workers, so they might tap into their collective experience for business advantage. Most firms now tap outside talent for scenario building and new product help.

Chapter 10

The Institutional Challenge: What to Reflect on?

Key Themes at a Glance

A belief that is slowly gaining acceptance — one that appears to be recognized by our competitors, but one that is still de-emphasized by academic mainstream thinkers — is that institutions matter. It is becoming increasingly obvious that an economy is more than a collection of companies.

Differences between countries' ideologies and social systems are becoming major factors in global competitiveness. The sound-bite centric, event-based Western news media tends to ignore this, except for reporting episodic crises and scandals. In fact, honest reporting of timely, relevant data is now itself a crucial infrastructure element for Information Age competitiveness.

Other key institutions are those that comprise and support the educational system. The scorecard on education indicates that America has major problems. Not only have other countries caught up, Americans are slipping further behind. That does not bode well for competitiveness.

The lessons from East Asia are clear: Subtle guidance works under special circumstances. Machine Age bureaucracy and Megastate government is generally harmful, but new infrastructure can help competitiveness.

> **New or renewed institutions can be developed to accommodate the requirements of an environment characterized by the Engines of Prosperity. These institutions will be grounded in ideologies, in the effectiveness of educational systems, and in new forms of interfirm collaboration.**
>
> **Government can be made responsive and accountable simply by returning to Constitutional basics. Educational infrastructure can be made effective, simply by de-politicizing educational institutions and measuring their effectiveness solely against the skills and competencies of the students they graduate.**
>
> **Whether we rebuild infrastructure and use institutions to our advantage — as in the case of East Asian competitors — or fail to capitalize on them for enhanced competitiveness constitutes the final challenge...**

Americans tend to distrust a central government, often for good reason. The model for the past 400 years has been the Western nation-state, invented by the French lawyer-politician Jean Bodin and described in his book *Six Livres de la Republique* (1576) in response to Spain's becoming wealthy and developing the first standing army since Roman times.[1]

Bodin prescribed such things as a centrally controlled civil service; central control of the military; a nationally controlled, professionally officered standing army; central control of coinage, taxes, and customs; and a centrally appointed professional judiciary. These revolutionary concepts opposed all that had existed in the thousand years since the fall of Rome.

Bodin's model appeared absurd in the late sixteenth century, an age of monarchies. The words "paradigm shift" were still centuries in the future, but that is what happened. Bodin's template was universally accepted as the external threats posed by ambitious and powerful states became clear. The choice was to either adapt or be conquered. Most adapted.

No country resisted this trend more than the United States.[2] It was started by a people whose rallying cry was "No taxation without representation," a people who valued their individual freedom over all else. For 11 years after independence was achieved, we allowed no central government. The former colonies instead operated under Articles of Confederation that allowed central government no enforcement powers and no rights to levy taxes. Our citizens were a fiercely independent and self-reliant people who knew a central government was not something to be trusted.

It was, ironically, not an external military threat, but local tyranny and economic weakness compared with the former English rule that finally drove the Confederation to become the nation-state that we now know as the United States. The U.S. Constitution was our founding father's effort to limit the powers of the new federal government.

Only extreme duress made the U.S. accept a central government. After years of depressed farm prices and abuses by state governments that harassed minorities (including the wealthy), confiscated property, taxed excessively, passed ex post facto laws, and issued floods of unsecured paper money, a national government started to look better. Ratification of the Constitution took another year and the promise that a bill of rights for individuals would be added.

Subsequent U.S. history is largely a story of distrust and adversarial conflict between the government and its citizens. Often it has been a tale of the federal government expanding its powers, especially the power to tax and control, at the expense of individuals, corporations, and the states.

The new government first tested its power by levying a "small excise tax on whisky." The result was public outrage. A tea tax from Britain had prompted rebellion, but a tax on whisky — even on a man's personal drinking whisky — was worse. A few tax collectors were killed, and this prompted the first use of federal troops against U.S. citizens, in 1790.

It gets little notice in high school history classes, but the Whisky Rebellion was a major event of the time. Whisky was a more common currency for trade than the worthless paper money of the day, and most farmers had stills. The government assembled 13,000 troops — an overwhelming force — led by George Washington himself, and threatened to "put to the sword" any who resisted the tax. No one did, so it was more of a political debate than a rebellion.

Similarly, one of the major issues, which caused the U.S. Civil War, was an intense disagreement between the North and the South over federal tariff policy. "King Cotton" was the most lucrative trade of the time, and tariffs — tax policy — decided who kept the profits.

The South wanted to export cotton to Europe, and the North wanted to protect its textile mills. These goals clashed, but the North had more clout in Congress. Though slavery was the moral high ground, Lincoln's policy to free slaves was not declared until well after the war started.

The massive spending and bloated bureaucracies we've grown accustomed to in the U.S. are relatively recent phenomena. Only in the past thirty years has federal control of schools and over social issues (e.g., bussing students, welfare) been allowed. The tax "reductions" of 1986 and 1993 were each the largest tax increases in history. Both administrations promised deficit reduction, but both lied. Neither reduced the deficit or even their spending levels.

In the United States, the effort to break up political power and the attempt to prevent the concentration of economic power have been seen as parallel steps toward liberty. The United States has a three-branch government because of fear that any one branch will become too dominant.

The great reformers in the American tradition have generally risen to strike down excessive concentrations of power, from Jefferson to Andrew Jackson to Teddy Roosevelt. People who have argued for centralizing and exercising power have generally had the excuse of wartime: Abraham Lincoln, Woodrow Wilson, Franklin D. Roosevelt, John F. Kennedy, and Lyndon Johnson.

The early 20th century saw many abuses of workers by companies, destructive speculation by Wall Street, and unfair competition by cartels. Excessive corporate power prompted government intervention and a raft of restrictive antitrust, securities, and banking laws.

The government's first significant foray against business was to prevent railroads from using their monopoly powers to gouge customers on freight rates. Their second intervention prevented Mr. Rockefeller from using his monopoly control of lighting oil to exploit renters. When capitalism collapsed in the 1930s, the federal government got involved in everything.

The Emergency Powers Act of 1933 vested unlimited power in the Executive Branch, wherein the President may seize property and commodities, organize and control production, regulate private enterprise, and restrict travel. This long forgotten act remains in effect today. Executive Orders and rulings and regulations from dozens of bureaucracies have the full force of law when printed in the Federal Register, and this despite the fact that the Constitution clearly says that the legislative branch should "make all laws."

Peter Drucker, the famous business theoretician, tracks the inexorable evolution of the nation-state into what he calls the Megastate in his best-selling book *Post-Capitalist Society*.[3] He says that by 1870 the nation-state model had triumphed everywhere, but it still functioned in a way that resembled Bodin's original model. From 1870 to 1970 these entities had mutated into a vastly different form, with a vastly different goal.

The nation-state had been formed to protect its citizens' lives and liberty from external threats, and its citizens' property from arbitrary acts of the sovereign. It was designed to be the guardian of civil society. The Megastate has a different goal. It is not the guardian, but the master of society. In its extreme totalitarian form, as exemplified by the former USSR, it replaced civil society completely, so that all society became political society.

Box 10-1

> **"The nation-state of 1970, a century later, bore little resemblance to Bodin's state or, indeed, to the nation-state of 1870. It had mutated into the *Megastate* — the same species perhaps as its 1870 progenitor, but as different from it as the Panther is from the pussycat."**
>
> Dr. Peter F. Drucker
> *Post-Capitalist Society*

Megastates, even in their least extreme, Anglo-American, form, consider a citizen's property to be held only at the discretion of the tax collector. It was World War I that made this transition possible, and the famous Austrian (later at Harvard) economist Joseph Schumpeter first noted it in his essay *Der Steuerstaat* (*The Fiscal State*, 1918).

The fiscal state enabled the Megastate. The first person to fully understand the social implications was not an academic or a politician, but a novelist. Franz Kafka's novels *The Trial* and *The Castle* (1926) are penetrating analyses of the Megastate.

Until World War I, no government in history could squeeze from its people more than a small percentage, perhaps five or six percent, of the country's national income. But World War I proved that every belligerent, even the poorest, could tax, increase deficits, and spend almost without limit. Spending soon exceeded total national incomes. Capital and assets obtained over decades or centuries was acquired and turned into war material in short order. The major belligerents came out of the war economically devastated.

Schumpeter pointed out the fundamental change.[4] For as long as governments had existed, the budget process had begun with the assessment of available revenues. Since "good causes" were limitless, the budget process consisted of knowing when and where to say no. These restraints made it impossible for governments to act as social or economic agencies.

Megastates know there is essentially no limit to available revenues. Such power is hard to resist. With the singular exception of Japan, which has lagged the trend, all the nation-states evolved into Megastates. Since governments no longer have to say no to anything, they can become involved in everything, especially social and economic policy.

Drucker's book starts with the fiscal state's ability to spend without practical limit.[5] Concurrently, the nation-state mutated into what Drucker calls "the nanny state," first invented by Bismark in the 1880s to combat the rising socialist tide. This is better called the Welfare or Entitlement State.

Box 10-2

> **"(Since World War II) all developed and many developing countries have become 'fiscal states.' They have all come to believe that there are no economic limits to what government can tax or borrow and, therefore, no economic limits to what government can spend."**
>
> Dr. Peter F. Drucker
> *Post-Capitalist Society*

Bismarck's actions were modest, but the principle was revolutionary. Government had been perceived as purely a political agency, but Bismark made it into a social agency by funding his own limited welfare measures — health insurance (sound familiar?), insurance against industrial accidents, and old-age pensions. Thirty years later, the British invented unemployment insurance. These early programs were limited, and the government's role was indirect.

The trend swept into the United States in 1932 with the election of Franklin Deleno Roosevelt, who instituted Social Security (1935) and the New Deal. In the bleak days of the Great Depression, Roosevelt's programs passed despite pockets of fierce resistance. There was nearly a Constitutional crisis over Social Security.

Americans resisted taxation. The U.S. had instituted a "temporary" income tax during the Civil War. It took a lifetime of struggle (until 1894) to have it declared unconstitutional. When memories had faded, the tax was resurrected to help pay for World War I (1913). The government again said it was temporary. They lied. The tax stayed and grew, eventually spawning the all-intrusive, ever-growing bureaucracy we know as the IRS.

Payroll taxes were started in 1939 to help fund Roosevelt's social programs. Today everything imaginable is taxed. Income is doubly or triply taxed if dividends, state and local taxes are considered. Some cities even charge fees for the rainwater that runs off a homeowner's roof.

After World War II, the state evolved from a provider to a manager, and this led to the Megastate as master of the economy. Today's economists agree on little, but nearly all of them consider the nation-state and its government the master of the national economy and the controller of its economic weather.

The next stage of evolution was into what Drucker calls The Cold War State. Government was now the master of the economy, so why wait for conflict to break out to defend one's people? That trend had also started in the 1890s, when the Germans in peacetime started building steel-clad ships to prepare for the next war.

Since war is inevitable, why not prepare early? Technology allows military advantage, if the weapons are available and the fighting men are trained before the outbreak of hostilities. The brevity of the Gulf War testifies to this.

There were attempts to resist the Cold War State during the thirties. Woodrow Wilson wanted the League of Nations to control national armaments, but it didn't work. Germany and Japan built military machines that almost took over the world, and few since World War II have tried to revert to "normal" peacetime economies. Nuclear weapons and the effectiveness of exotic technology on the battlefield have made the condition permanent.

So the nation-state mutated into ever more intrusive forms. First into the Fiscal State, then to the Welfare State, and finally into the Cold War State. The Cold War State has become economically destructive, diverting money and precious talent into economically unproductive defense work. Drucker says there is little doubt that Latin America, rather than East Asia, would have been the economic miracle of the 1960s and 1970s, had it not wasted all its money and trained people building up huge armed forces.

Lyndon Johnson, a consummate politician, was the first to combine the forms, offering both Cold War and Welfare with his "guns and butter" policies. This led to colossal deficits and finally to the next step in the mutation, to what Drucker calls the Pork Barrel State, where government spending becomes the way for politicians to buy votes and stay in power. With this step, the Megastate was complete. While citizens were once served by the nation-state, the Megastate is their master, the center of power.

Box 10-3

"Democratic government rests on the belief that the first job of elected representatives is to defend their constituents against rapacious government. The pork barrel state thus increasingly undermines the foundations of a free society. The elected representatives fleece their constituents to enrich special-interest groups and thereby to buy their votes. This is a denial of the concept of citizenship — and is beginning to be seen as such."

Dr. Peter F. Drucker
Post-Capitalist Society

Ever notice that the Megastate's bureaucrats devise programs that are endless and spend money on all sides of every issue? Funding is provided to grow tobacco, to warn consumers of the hazards of the weed, and to research less harmful strains.

The U.S. spends some $12 billion each year on the war on drugs, and also spends $1.4 billion through Supplemental Security Income (SSI), mostly to buy drugs and alcohol for addicts. One Denver liquor storeowner got $160,000 annually on behalf of 40 alcoholics he supplied with booze. (In theory, addicts were to get treatment. In 1993, only 8% were confirmed as getting treatment. The cure rate was near zero, because if an addict was cured or worked his funding was cut off.)[6]

The Pentagon and Defense Department spend $2.5 billion per year on travel. The government spends another $2 billion to audit these expenses.[7]

The same high-friction, low-output phenomena are evident at state and local levels, and the powers of regulatory and enforcement branches are ever increasing. Customs can "inspect" imports with chain saws with impunity. If property is seized by the drug police or IRS, the owner must acquiesce or post bond, hire lawyers, and somehow prove his goods are innocent.

"Has the Megastate worked?" asks Drucker. It certainly did not in its totalitarian Nazi or Communist forms, but what of its more moderate embodiments? Has it worked in the developed countries of Western Europe or the United States? Drucker's conclusion, "Hardly any better."

To move into the future, the first step, paradoxically, is to turn back the clock to Constitutional government. Major segments of our society have become totally dependent on government control and funding. Government of the people, by the people and for the people and been eroded. The present form, government of the people, by the Megastate and for the Megastate's officials and bureaucrats can be reversed if citizens, educators, and corporations all focus and work together.

Asian Alternatives

Drucker's most helpful comment is that Japan is the single major exception to the Megastate model.[8] Ironically, it has been the country in which the government has played an indirect, highly restricted, and restrained role. Japan in many ways looks like an 1890s economy, in that government service enjoys tremendous respect and government works closely with big business to accomplish national objectives.[9]

Box 10-4

> **"The Japanese government has never picked winners and losers. Their strategies have always been bottom-up, industry led strategies, where government was always a participant but never a dictator."**
>
> Dr. Lester Thurow
> *Head to Head*

Japan has almost no national social programs, except for national health insurance, imposed on it by the United States during the occupation.[10] It has so far avoided being a Cold War State. Its role for government is that of a guardian or a wise parent, not a manager or a master. (Its government's actions foster becoming healthy rather than feeling good.) Whether this can continue given Japan's prosperity is an open question, but, so far, it is the most notable exception to the Megastate model.

Most would agree that the United States' version of the Megastate is a declining society. The famous Japanese novelist, Akiyuki Nosaka says watching the U.S. is like "a test run for the decline of the human race." They have a new word, *kenbei,* which means a gut-level dislike for America. Many Japanese see us as a society plagued by illiteracy, crime, poverty, drugs, and corruption. Some are sad and nostalgic, but many are contemptuous.

The economic systems that are showing the best results in the last decade of the twentieth century have been those of the Pacific Rim, especially Japan. These come about from vastly different viewpoints, backgrounds and cultures. Despite Japan's much publicized "disarray," its trade balance and economic power continues to grow.

Pacific Rim and Asian national approaches to business are considerably different from Western practices. Japan had the opportunity, with U.S. help, to take the best things they could find from U.S. government and industry practice. Then they merged these with the best practices from their own and other cultures when they rebuilt after World War II.

As Lester Thurow has said, the contest in coming years will test these two different forms of capitalism against each other.[11] Cold war has turned to economic war.

To Westerners, economics is a game from which some or all can emerge as winners. Conversely, Asian history instructs many Koreans, Chinese, and Japanese that economic competition is a form of war.[12] To be strong is much better than to be weak, and giving orders is better than taking them.

Asians think the way to be strong, to give orders, to have independence and control — to win — is to keep in mind the difference between "us" and "them."[13] This perspective comes naturally to Koreans when thinking about Japan, or to Canadians when thinking about the United States, or to the French when thinking about almost anyone. It does not come naturally to most Americans.

There are more unfair trade examples from Japan than from the other countries, but only because Japan got there first. Korea, for instance, would love to be just as nationalistic, but, so far, they are still too weak.

Koreans worry about being taken advantage of, as the Japanese have done.[14] That fear makes them tough business negotiators. Business negotiations with Korean firms invariably involve their government and pressure for eventual royalty free technology rights. The U.S. practices individualistic capitalism, while the Pacific Rim or Asian countries practice collaborative capitalism. Some economists call the Asian model "structural mercantilism."

Whatever one calls it, the most dynamic Asian countries all use government-industry links, subsidized capital, large conglomerates, and resistance to imports as trade weapons. Europe is somewhere between, trying mightily to move from narrow local concerns toward becoming a collaborative, closely integrated, multinational-trading bloc.

Pacific Rim practices come from survival needs. Japan was devastated by World War II, desperate, and had no natural assets for rebuilding. Its only resource was people, so Japan used it well. Their model for prosperity is to have government, citizens, workers, and companies pull together in harmony. This worked.

Korea is an even better example of a devastated and impoverished country, which focused its people to create prosperity. Korea has been growing its economy at twice the rate of Japan's.

The method in countries favoring collaborative capitalism is for all stakeholders (e.g., management, employees, banks, investors, communities, government, and suppliers) to work together toward a goal of mutual prosperity. The systems are adaptive, subtle, and often covert. They are based more on relationships and mutual trust than upon law or formal rules.

The touchstone of collaborative capitalism is to do whatever works best for the mutual good. Harmony and trust are emphasized. The role of the leader is guidance, wisdom, and help. Guidance is not only acceptable, but also wanted. Mutual good to the Japanese means high wage jobs and exports, not abstract economic principles or trade theory.

Insiders are treated with honor and integrity, but outsiders may be exploited.

In the U.S. model, the government's basic reason for having an economy is to raise the consumer's standard of living. In the Asian model an economy's main purpose is to increase the collective national strength. Ideally, the goal is to make the nation independent and self-sufficient, so that it does not rely on outsiders for its survival.

The U.S. goal is materialistic and economic. Each business fends for itself, and the government attempts to redistribute what wealth results. The Asian-style goal is political, and it comes from the long experience of being oppressed by people with stronger economies and technologies.

U.S. ideology views concentrated power as an evil. Therefore it has developed elaborate schemes for dividing and breaking up power when it becomes concentrated. Asian collaborative capitalism views concentrated power as a fact of life. It has developed elaborate systems for ensuring that the power is used for the long-term national good.

The U.S. model views surprise as a fact of economic life. We believe that it is precisely because markets are fluid and unpredictable that they work. The Asian-style system deeply mistrusts markets, so governments intervene to lower the risk for their firms.

The Pacific Rim sees competition as a useful tool for keeping companies on their toes, but not as a way to resolve any of the major questions of life. It would <u>never</u> consider using the free market to decide how a society should be run, or in what direction its economy should unfold.

This can be best understood by analogy to the U.S. military. In the U.S. the Army competes with the Navy and other services for funds. This rivalry, in theory, keeps all the services sharp. But the services don't cast votes or place bids to decide when and where the nation should fight. In similar manner, Asians don't leave economic policy decisions to a market or to their corporations.

Asian business practice is a very different approach from the profit seeking, adversarial, law-based systems used in the West. Japanese and Korean corporations willingly accept artificially low profits in return for full employment and increasing their world market share.

Korea is more direct about protecting its interests than the Japanese. The 1988 Seoul Olympics did for the country what the 1964 Tokyo Olympics had done for Japan. Anything seemed possible. In the fashionable parts of Seoul, young women wore miniskirts and young men hung out all night. Koreans were becoming Westernized.

By 1990 Korea's trade surplus was slumping. So the government gave some guidance to correct matters, an "anti-luxury" campaign. The national tax office announced that "extravagance beyond one's reported means" would invite tax scrutiny.

In effect this meant that anyone who bought a foreign car, a Mercury Sable, Lincoln Continental, Mercedes, or BMW could expect to be put through the tax wringer. This was a more serious threat in Korea than in most other countries, because so much business is off the books.

Tariffs and other barriers had already raised the price of these cars to more than twice what they would cost in the United States. That hadn't choked off sales, but the tax threat did. Sales of the Sable virtually stopped after the taxmen stepped in.

This is very much an us-versus-them model of the world. People everywhere are xenophobic and clannish, but in the Western model this is thought to be a lamentable, surmountable failing. The Asian model assumes that it is a natural and permanent condition. The world consists of us and them, and no one else will look out for us.

Critics note that Japan's system is insider-oriented and gift-based, and, therefore, more subject to corruption. Still, the underpinnings of Japan's system come more from mutual trust and shared objectives than from gifts. Both systems have control mechanisms. Where we attempt to rectify breach of law, they punish violation of trust. Most Japanese see the U.S. as more corrupt than their own society, but the point is arguable.

In the U.S. we put our trust in rules and laws rather than people, because people can become corrupted, racist, or bigoted. Unfortunately, the Megastate has twisted representative government into a system whereby government takes "donations" or "political contributions" under the paper thin pretext that these are technically legal and "can't be proven" to have influenced behavior.

Campaign reform is often discussed. It is a great charade: one year the Senate passes reform, but the House doesn't manage to get around to it. Next time they trade places. Suffice it to say that we have corruption problems too, and that we need to deal with them as an urgent matter.

Still, the deepest criticism of Japanese politics, made by the Dutch writer Karel van Wolferen, is that it lacks a definable center of political accountability. In the French or American system, a President must finally make big choices, whereas in the Japanese system, he says, the buck never stops anywhere. What critics fail to understand is that this is exactly what gives the Japanese system such phenomenal effectiveness.

The classic illustration of this "problem" is Japan's apparent paralysis during the first month after Iraq invaded Kuwait. The issue was what Japan should contribute. The standard critique outside Japan was that the country was not doing its "fair share."

That critique was superficial. Japan moved cautiously, because the decision was important. They need oil and care little about the moral rights of the Kuwaitis. The last thing Japan wanted was a situation which left the U.S. with control of Kuwait or Iraq oil.

It may be hard to accept, especially for U.S. liberals, but it was well known to the oil industry, and certainly to the Japanese — who do their homework meticulously and get 60% of their oil from that region, that oil would have been *cheaper* if Iraq controlled Kuwait's wells.

Kuwait and Saudi Arabia have almost no native population and no urgent need for immediate income, so they want to keep production low. They want to keep prices high.

Iraq, with severe overpopulation and no other valuable natural resources, needs to sell as much oil as it can. It must sell oil, and doesn't care if that causes world prices to drop. No wonder Saddam Hussein so badly miscalculated the world's reaction to his terrorist act.

The startling thing to those who expected "normal" Megastate behavior was that Japan seemed incapable of deciding what its position was. The Foreign Ministry announced one policy, but the Finance Ministry disavowed it. The Prime Minister at the time, Toshiki Kaifu, was scheduled to go on a trip to the Middle East, but officials in the Foreign Ministry called the trip off. So what? Japan eventually contributed enough money to avoid criticism, avoided risk, and accomplished all its major objectives.

One of the authors took a course in decision methods from Yale's noted Vic Vroom. His research was not into what specific decisions were best, but what decision processes were optimal under certain conditions. If the quality of the decision is paramount, what process gives the best results?

Vroom's research proved that for one type of problem — uncertain, knowledge-intensive situations — distributed, consensus-based processes yield higher quality decisions. One experiment he has run many times is to form people into groups, give them poorly understood problems, and compare the individual to the group consensus decisions. He consistently found that consensus decisions were much better than the average of the individual decisions made within the group. Quite often they were of better quality than the best individual decisions within the group.

Box 10-5

Note: For other types of problems, other processes may be better. For example, autocratic decision processes work well for clear choices or trivial problems.

Japan's diffused system of influence has the best track record in the world in modern business success. Western management and government control systems resemble centralized data processing, but Japan's is more like a distributed computer network. While we decide and control at the federal government level, through bureaucratic procedure, their system has local intelligence and decision making ability.

Bureaucrats and politicians make U.S. industrial policy decisions publicly in Washington. Accepted methods are based on clout, popularity, emotion, lobbying, pork barrel politics, and sound bites. Megastates are not noted for deep thought or reasoned consensus.

Conversely, no outsider is ever really sure where Japan's decisions are made. We are sure that the preferred decision process is collaboration, based on knowledge, expertise, and experience. We do know the Japanese characterize our centralized system as "loser-driven," since the main impact of political lobbying is usually protection for dying industries.

Bad central decisions cannot disrupt Japan's industrial policies. Theirs is also a convenient system when they wish to say "no," as in discussions over market access. Japan can indeed say "no," and they do it very well. They can quickly fire their prime ministers to thwart Western pressure. Japan's government is easily disposable if it fails to serve the national good.

Rethinking Subtle Guidance

When Japan's trade negotiators regretfully claim that they are powerless to do more to open their markets, the statement is true. It is also false. The same is true when they say "yes" about opening markets. It is like patting us on the head and saying, "Yeah, yeah." They know they can stall. There will soon be a new administration in Washington, and negotiations will start over.

It is interesting that diffused influence allows more precise implementation than central control. One of the authors was in Japan during a railroad "strike." Every train was ten minutes late — not nine, eleven, but exactly ten minutes. The trade unions would not do anything destructive, but they wanted to signal a grievance and this was the method.

In general, although most other Asian governments do have centers of power, collaborative capitalism favors gentle but firm "guidance." It is not acceptable to cause societal harm to get one's way under collaborative capitalism. Selfish interests are to be subordinated to the greater good.

Whether the center of politics is weak, as in Japan, or strong, as everywhere else, the political system as a whole has generally been authoritarian in Asia. Compared with Western societies, and especially the Anglo-American system, Asian states have been less embarrassed and more explicit about the government's role in shaping society.

The contrast is obviously sharper with America than with, say, France, which operates a Japanese-style system, but without the social control. The Japanese system also resembles the most successful parts of government-business interactions in the United States, such as weapons design or medical research.

A darker example of how collaborative capitalism works in Japan is the famous Lion Petroleum case. Lion is a small Japanese refinery that in 1984 tried to import gasoline from Singapore for sale in its Tokyo gas stations. MITI nurtures the Japanese refining industry, carefully protecting it from foreign competition. Japan keeps fuel costs low and consumer gasoline prices high.

Lion wanted to profit by selling inexpensive imported gasoline — a logical strategy. This was allowed. It was perfectly legal to import gasoline. There were no tariffs, no quotas, and no import restrictions. It looked like an open market opportunity, but, of course, it was not.

For Lion to import gasoline, all it had to do was notify MITI. Unfortunately, MITI refused to accept the notification. While never issuing a direct order, MITI's guidance made it clear that Lion and its backers were not to import the gasoline.

To prevent others from having notions of undercutting established pricing; the Japanese government took action. Their representatives in Singapore and the Philippines warned those governments against allowing sales to gasoline importers.

It is revealing that Lion itself made no objection. They certainly did not try official protest or litigation with the government. That is how Japan's system works, and Lion understands the unwritten rules.

The Lion case is an old example, but nothing has changed. Japan puts its economic strength over consumers' interests, and it always has. This is so alien to the U.S., which puts consumer interests first, that we have a very hard time understanding that their rules are different.

In 1988-90, Japan's government protected its market from a proven Merck Corporation product called MMR. Merck's product was trademarked, patented, and extremely safe, having been used on more than 100 million children without serious side effects. MMR is used to protect children against measles, mumps, and rubella (German measles) with one inoculation.

Merck Corporation was number one on *Fortune* Magazine's "American's Most Admired Corporations" from 1986 to 1991. This is the same spot IBM held before computers, software and electronics became targeted industries. Merck, now targeted, lost this honor to Rubbermaid after laying off some 10,000 employees in 1993, falling to 11th place. (IBM, in comparison, was 354th and in trouble.)

The U.S. still leads in drugs, but defenses are thin. Product liability issues often push all but the dominant supplier out of the market. U.S. consumers rely on single firms to supply vaccines for polio, measles, mumps, rubella, and rabies. (Source: *National Center for Manufacturing Sciences*, December, 1993.)

To promote the growth of Japan's pharmaceutical industry, and to avoid using Merck's product, the Japanese government asked each of three companies to produce its best vaccine for one of the diseases covered by MMR. The government combined these into a new vaccine, which it also called MMR.

When the vaccine was ready, in early 1989, the Ministry of Health and Welfare began a mandatory nationwide inoculation program for children. "Rather than use foreign products, we wanted Japanese products because they are of better quality," an official of Japan's Association of Biological Manufacturers told Leslie Helm, who reported the story in the Los Angeles Times.

In fact, Japan knew its MMR drug was of much worse quality than the foreign alternative. They reportedly expected their drug would produce side effects in one out of 100,000 inoculations, a thousand-fold worse result than with Merck's proven product.

The actual incidence of side effects was at least 100 times worse than expected. The most serious side effects were meningitis and encephalitis, which killed some children and left others paralyzed or brain-damaged.

By the end of 1989, the government had made the Japanese MMR vaccine optional rather than mandatory, but it left the vaccine on the market until remaining stocks were consumed. It still did not approve the safer Merck product for sale in Japan. Based on past actions, it never will unless a Japanese firm can buy Merck or secure a technology license.

Japanese doctors have now returned to giving separate immunizations for the three diseases. Biotechnology is one area favored by the Engines of Prosperity, and Japan can couple guidance to regulation when they deem it in their national interest. The Merck example is persuasive for the war metaphor. The Japanese love their children and usually give them much better care than we do our own. Only the highest national priorities — like an economic war — could take precedence.

Guidance is a totally different tool than regulation. The U.S. Department of Commerce would never issue guidance to our semiconductor industry to hold down production. If they did, resistance would be fierce and litigation would be certain. But the Pacific Rim economies operate differently, their governments take different prerogatives, and their citizens have different expectations.

Psychologists say that one measure of maturity is the delay between wanting a thing and reaching for it. Under collaborative capitalism industries often accept things that cause them pain, if they trust that it serves a larger purpose and that they will eventually benefit.

It is unlikely that Japanese electronics executives are happy about being forced to buy U.S. semiconductors — the much discussed 20% target, which MITI reluctantly agreed to. They also know that this (possible) concession is in their nation's and industry's best interests, and probably temporary. (They were right. The targets were removed as this book was being written.)

Companies don't have to obey MITI. There are rare examples where guidance was refused, though usually at a price. Honda, for example, was guided not to enter the automobile industry and Sony was guided to license its Beta format tapes.

Pacific Rim countries think they can't really go wrong by giving consumers too little, but they can easily go wrong by giving them too much. During the collapse of Japan's bubble economy, in 1991 and 1992, government officials said privately that an atmosphere of hardship was useful. Consumerism had been getting out of hand, and the bubble's collapse would have a tonic effect, and without imposing real hardship on Japan or endangering Japan's long-term prospects.

Business-failure rates among Japanese manufacturing and construction firms were lower during the "crash" years of the early 1990s than they averaged during the booming 1980s.

Ideology Versus Reality

Our diplomatic muddle over trade is the result of ideological fixation. The orthodox U.S. trade theorists insist that laissez-faire is the best route to economic growth. Confronted by the success of nations like Japan and Korea, they conclude that the resulting trade deficit must reflect some internal failure, like not working hard enough. Japanese propaganda, of course, plays on this.

It is simply not true. Whatever problems we have, our workers are still trying. U.S. real labor cost is low and U.S. male workers work more hours weekly compared to the Japanese. (Sorry, ladies. Females do not yet play a significant role in Japan's work force.)

In the end, believing strongly in laissez-faire and open markets, our politicians conclude that the only remedy is for the U.S. to set a good example by not fighting back. Our trade policy is a constant refrain of "lower tariffs" and a sequence of free-trade blocs — surely a contradiction in terms — whose members piously promise to lower tariffs for those in the bloc.

There have been only two eras where free trade is known to have worked as an economic policy. These were during the times when Britain totally dominated world trade, and when the U.S. was the only major trading nation left undamaged by World War II.

The GATT-Bretton Woods trading system was adopted in 1945 at the urging of the United States. America believed that it could not be prosperous unless the world was prosperous and everyone had equal access to markets and raw materials.

Our free trade concept was a great success in rebuilding the world's economy. It worked, and, by working, changed the world. The U.S. in 1945 was richer than the rest of the world combined, but by the late 1980s it was only one of several somewhat wealthy countries. Its trajectory has been downwards, while other economies have been growing.

Unfortunately, economic practices designed for a unipolar world don't work well in a world of coequal trading partners. A growing world economy no longer means a growing U.S. economy, and the days of effortless U.S. prosperity are over.

The U.S. is still locked into the old patterns of behavior. In exchange for tariff cuts and free-trade blocs, the U.S. typically concedes the issues that matter the most to U.S. exporters — such as limits to foreign subsidies, intellectual property protection, and reciprocal market access. We totally fail to understand that Japan is not about to change its economic system to get a greater access to the U.S. market. It already has all the access it needs.

As we stand on the threshold of the 21st century, both the collaborative capitalism of the Pacific Rim and the individualistic capitalism of the West face significant challenges. The Engines of Prosperity are churning, and competition is head-to-head and intense. Those who can ride the waves of change will prosper, but even minor errors or delays can leave firms and nations in stagnant economic backwaters where margins are thin and overcapacity is abundant.

In 1776, our founding fathers did their best to preserve freedom and block direct federal government intervention with people or businesses. They did a wonderful job, but centuries of relentless pressure have eroded the barriers they built with such determination and care. The U.S. has moved to ever more pervasive and ever higher levels of government control, involvement, and entitlement. This hasn't been working, so pressure for change is building.

All organisms, when stressed, take the actions that helped the last time. As prosperity declines, the Megastate presses for more control, more free trade, more social programs, and lower tariffs.

Regulation, litigation, and a lack of focus and infrastructure make it difficult to do business in the U.S. These things impede prosperity, but are outside the scope of U.S. economic models. None of the experts see solutions or predict a near-term economic recovery, and most are riding the free trade, laissez-faire bandwagon. We've been told the recession is over and the economy is turning around for so long that few, especially those without jobs, believe it.

Just as we will see that addressing product development issues in business requires breaking down turf and changing behavior, so it is with the government. Meaningful solutions require "outside the box" thinking and crossing functional boundaries, and we are ill equipped to do that. Who dares to oppose a bandwagon of impeccably credentialed economists and powerful politicians?

We are, of course, not the only ones with troubles. The Japanese economic system worked superbly until about 1991, but now it too is stalled and under considerable stress. Their model for success has been to adapt Western technology to create better and cheaper products. Having taken over entire industries — e.g., consumer electronics — leaves few Western innovators for Japan to copy and out-implement. This is becoming a major problem for them.

That means that Japan's collaborative system must develop greatly increased access to Western inventiveness (they are trying) or develop this resource itself. Unfortunately, innovation is not a strength in their society. In addition, international resistance to exploitative Japanese trade practices is mounting and the Yen is 250% stronger than when the Japanese first flooded Western markets with products.

Europe must overcome centuries of national conflicts to unify as a trade block. They seem determined to accomplish this. They must also solve some mysteries to fully participate in industries with *steep learning curves* and *demand amplification*, the dynamo markets. Europe has been successful in chemicals, machine tools, and aircraft, but, despite repeated and massive efforts, they have been dismal failures in electronics, semiconductors, and software. No one is sure why.

In the U.S. the slogan "It's the economy, stupid!" persists, but the future is unclear. Will the U.S. move toward a form of Pacific Rim collaborative capitalism? Can the U.S. government accept a role of guidance versus bureaucratic control? If so, how could this be done under the present political and legal system? If not, what model will we use to compete in global trade?

For a first step, what can be done to stop foreign lobbyists from buying the very people who make the laws and policies that govern us? Who will be that cat? Will the U.S. continue to ignore its difficult decisions?

This is a branch point in history. Depending on if and how America's global competitiveness problems are resolved, vastly different futures emerge. There is a strong possibility that solving this problem may cause others to fall into place. Return to a traditionally American "abundance" world view where people can shape their own lives positively has much to recommend it over a scarcity "it's all awful" view, where the only hope is the Megastate. The classic Machine Age "go West, young man" counsel could evolve into Information Age "gain knowledge, skills, and wisdom — so you can prosper."

With present levels of uncertainty and tension, no one can know what the future holds. Still, the authors are inclined to think economic prosperity will remain the central issue, since the availability of family wage jobs forms the foundation for addressing all the other issues. We think that the U.S. will, by the end of the decade, either make radical changes in its economic models and policies or overtly retreat from its role as a major world power.

In the end, the Engines of Prosperity will drive change and reform. The end of the cold war will change Western capitalism more than it changed Eastern Europe. We can choose protecting the past or embracing the future, but we will change in major ways.

Rethinking Free Trade Ideologies

Depression era U.S. economic theories and models assumed static markets and adversarial relationships between business, workers, and government. These models, and the government's perceived need to control, are still with us, though the world is now greatly changed.

The U.S. government mostly views business as a zero sum game. Government assumes that for every winner there must be a loser, but the Engines of Prosperity have changed that.

During the early 20th century markets were based on heavy industry or agriculture. They grew slowly, if at all. The economic planners of the era did not even imagine, much less consider, scenarios where all parties — labor, industry, and government — could win.

Today we carry this legacy. The irony of protecting people from the employers who provide jobs and perhaps purpose is ignored, as is the danger of excessive regulation causing both the workers and the companies that employ them to fail. Indeed, nowhere in the world is it easier to sell companies to foreign interests, or cheaper to layoff workers.

This narrow view of the world allows no possibility of technology-driven market growth allowing mutual prosperity. It is a binary win/lose viewpoint that assumes, since wealth is constrained, only the distribution of wealth matters.

Washington fears corporations will win and gain all the wealth, thus exploiting workers. Preventing such abuse and exploitation justifies government intervention. Even when Washington sees an industry growing, hiring workers, and raising incomes at a high rate, it becomes fearful. All the industrial age economic models predict such abnormal events drive inflation and eventual collapse. That too justifies government intervention.

Other U.S. economic assumptions are that, since all markets have the same dynamics, it does not matter what business areas are pursued. The government's proper role is seen as a mix of control to protect the public, occasional subsidies to prevent strategically critical bankruptcies (e.g., Lockheed or Chrysler), and nineteenth century laissez-faire capitalism.

Overlaid on this worldview is a cold-war-inspired government role of funding technology creation for defense applications, and occasional "big science" initiatives like the space program and the super collider. Defense interests and old-line industries are much more adept at lobbying Congress than "sunrise" industries based on commercial application of technology. Few commercial high tech firms — and virtually no new ventures that drive high wage job creation — maintain Washington offices.

The key trade policies advocated by the U.S. are free trade, low tariffs, and open markets. On top of this, the U.S. government structure protects dying industries and encourages "pork barrel" politics. Special interests "invest in" politicians to secure favors. Politicians use the money to finance their campaigns. The special interests recoup their money when legislation, government programs, or public spending is diverted, as they desire.

There is disturbing evidence of foreign attempts to secure trade advantage and military technology. Senator Fred Thompson, who chaired 1997 Senate hearings on donations said that his contentions of foreign influence "have been confirmed by the highest levels of U.S. security."[15] Strange "coincidences" abound, like the Commerce Department's disadvantageous agreements with Japan to weaken the U.S. patent system.[16] Senator Thompson's hearings eventually fizzled out, after much acrimony and with little result.

Have we reverted to taxation without representation? Many citizens think so. Members of Congress freely and legally sell themselves to foreign interests to secure needed campaign funds. Gallup polls say that "those who trust Washington to do the right thing all or most of the time" has declined from almost 80% in the 60s to about 20% in the 90s. That is a total flip-flop and a strong signal of crisis.[17]

Barbarians at the Keyboard

If advantage is nonexistent or fleeting in natural resources, capital, and technology, what is left? Workforce skills now drive the competitiveness of both nations and businesses. This area has the greatest potential for sustainable competitive advantage in this decade. The old model of "flesh robots" that do what they are told, is shifting to a model of empowered, savvy, self-managing work teams.

Countries with high workforce skills will be high wage, prosperous nations. Countries with low skills will work for third world labor rates, and their citizens will neither live long nor prosper.

Right now, the United States workforce still has good skills.[18] Those in the workforce are highly productive, and adequately trained. In 1993 we had the most productive workers in the world on an hourly cost basis. Unfortunately, our economy is stalled or declining for other reasons, described below, and time is running out.

Perhaps the greatest problem the U.S. faces in the future is deteriorating workforce quality.[19] Estimates of illiteracy in the work force range as high as 40%. There are now three generations of inner-city poor who have grown up without anyone in the family ever having had a legal job.

Social issues also sap vitality. One family in six is touched by a major crime each year. One birth in four is illegitimate, and over one child in nine is raised on welfare. We have a higher percentage of our population in prison than any society. Schools have become dangerous places.

The Japanese say that the Gross National Product of nations will track the ability of their workforce to be reeducated and retrained.[20] Unfortunately, in the U.S., as disposable income drops and jobs become scarce, it is harder for individuals to fund retraining.

Box 10-6

> **"Half of all U.S. manufacturing companies spend less than one percent of annual budgets on the enhancement of workforce skills."**
>
> National Center for Manufacturing Sciences, 1993

Companies are now so lean and mean that, with exceptions, they have cut worker training.[21] Unemployment will pay for a while if you look for employment at a low wage job, but it will usually not pay you to go to school so you can get a good job. This makes it very difficult for middle-aged people who have spent their lives in, say, defense, automobile, or timber industries to recover from layoffs.

A four-year Department of Education study gave 35–40 tests to some 26,000 people who were a cross section of the population.[22] Their 150-page 1993 report concluded about half the U.S. workforce over age 16 are unfit for any jobs requiring basic reading and writing.

It is hard to believe those results, but the trend is disturbing. A study in the late 80s claimed a 37% functional illiteracy rate in the U.S. workforce. SAT scores dropped 76 points between 1960 and 1990.

Ranking the U.S. against other industrialized nations statistically, yields a disturbing pattern. For scientists and technicians per 1,000 people, we are #29 of 30. For books published per 1,000 people, we are #26 of 26.

For the longest school year in days, we are #15 of 16. For population percentage covered by health care, we are #19 of 19. For growth of exports (nine-year average) we are #19 of 19. For science test scores (14-year olds) we are #15 of 17.

U.S. secondary school children work 1,000 hours per year. Japanese children put in 1,600 hours per year. At the senior high school level the average U.S. student does 3.8 hours of homework per week, but the Japanese student does 19.0 hours.

American adults did spend more time with their families and less time watching TV than their Japanese equivalent (16 hours versus 7, and 12 versus 20). There have also been several studies that we spend at or near the top of the industrialized world in the cost to educate K-12 students, though we are #8 of 8 nations in what we pay teachers. (A good tabulation of such data is Shapiro's *We're Number One*, by Vintage Books.)[23]

Why does this matter in a book about modern management and business prosperity? Because the new models for work depend on having an advanced, trustworthy, and educated workforce. Barbarians will get third world incomes and lifestyles. They will not participate in the dynamo industries, except perhaps to work in foreign owned screwdriver plants for low wages.

Experts differ strongly about how we can best educate our children, so there may be more than one "best" way. That doesn't matter. If we can agree that what we are doing now is non-competitive, there are many very effective educational systems around the world that can be copied. Why not test several alternatives?

Box 10-7

"If you ask what is the good of education, the answer is easy — that education makes good men, and that good men act nobly."

Plato

Carver Mead, a famous educator himself, once commented that simply getting our public schools back from bureaucratic control would solve most problems. Through the ages, the West's wisest counsel has been to focus on the ethics, integrity, dignity, and competency of our citizens.

The need for education, of course, goes far beyond K-12 schooling. In the United States, only about 10% of young workers receive any formal company training compared to more than half in Japan and Germany. Most U.S. companies don't bother to train because they view workers as disposable.

Asian success is often attributed to Zen Buddhism as practiced by warriors, and Miyamoto Musashi's *A Book of Five Rings* is a popular reading for Japanese and some Western managers.[24] Paraphrased, the advice is simple: total honesty, broad and deep training, intuitive judgment and understanding, focus on gainful actions, close attention to subtlety and details, and avoiding all waste.

Towards the New Enlightened Workforce

Probably the deepest philosophical change needed for management trained Westerners, is the acceptance that we have come up against a more advanced technology. This is not made of silicon or rare metals, but rather it is a technology for how people train and work together as effective teams.

For decades Western educators and consultants have developed process after process to get more output from existing workers and organizations. Think of a farmer who still uses mules to pull his plow. No matter what training he gives the mules, he can't produce the results of his competitors using more advanced technology.

There is a new technology changing the world: enlightened, empowered workers. In the proper environments, this is providing breakthrough results. What is the proper environment?

It starts with total honesty. What are our goals, what are our resources, and what are our limits? We need to perceive the world clearly and honestly, but most Western corporations twist reality to serve internal purposes and to preserve their existing power structures. The market tends to sort that out eventually, but "spin doctors" are also common in government. We now apparently accept tax dollars being diverted to improve the images of the bureaucracies spending the money.

Misinformation or disinformation is a problem. If one thinks dishonestly, then one cannot be honest with himself. Without self-honesty, clear thought is not possible. Without clear thought, action is misguided.

Box 10-8

"The biggest trend that I see? The sophistication of deception is increasing at a greater rate than the technology for verification. That means the end of truth."

Alvin Toffler, Futurist, October 1996

We need both honest data and total, broad, endless training. As much as is possible, workers must know every art and every profession. Without training, one cannot act with skill. Action without skill or knowledge is unlikely to be effective.

What of educational and social resource decisions? Should we invest so that our talented and gifted children can lead the world? Or should we instead invest to help the disadvantaged? The classic Megastate answer, "do both," is now inoperative. That only works when there is enough resource to do both. Clearly, there are not enough resources to do everything. Clearly, throwing more money at existing bureaucracies is unlikely to help and might even make things worse.

Mostly the U.S. moves resource to problems, while our trading partners do the opposite — investing in opportunity. Germany and Japan are doing almost ten times better at developing exceptional students. *Wall Street Journal* reported that only 4% of U.S. 18-year-olds pass advanced placement tests in high school subjects. In contrast, the figure for both Japan and Germany is 36%.[25]

Subtle, smart government, working closely with business and citizens, can help point our economy toward growing commercial markets. Asian firms have moved from steel to electronics, from tractors to computers, and from ball bearings to semiconductors.

Also — and this is another big change for management — intuitive ability in the heat of action is often the crucial difference between success and failure. Intuition and perception are as much to be trained and practiced as are analytical abilities. Every senior military commander knows this, but we have largely forgotten it in business.

Intuition depends on experience, so seasoned workers are needed. Unfortunately, these are exactly the ones being discarded.

Box 10-9

"What we have left (after downsizing) are parade ground officers. OK for barracks duty, but useless in combat."

A DuPont Executive, 1994

The biggest difference in the Information Age is that today's important action usually occurs within a fast changing, complex, uncertain, holistic environment. There is no time to ponder and analyze on this battlefield. Intuition and fast action is necessary.

Fierce, toe-to-toe, technology-based, market competition is a complex, contradictory situation of flux and motion. Totalities must be instantly synthesized and understood. Much of the information flooding in to a modern manager is incomplete, inaccurate, or distorted, some made deliberately so by their adversaries. This means that intuition and perception are critical cognitive functions that must be trained and exercised long before the battle begins.

Competent people who have trained and practiced for years are invaluable assets, not costs to be cut. These people are a corporation's and a nation's major strategic assets in the Information Age.

Then there is the need for subtle understanding and close attention to detail. Repeatedly, things like minor software "bugs" have destroyed corporate credibility or wasted enormous resources. The most trivial components have resulted in redesign, litigation, and the recalls of fleets of vehicles.

The need for detail is embedded in Western cautionary tales — "for the want of a nail." Still, the fear-based, do-what-you-are-told styles of management rarely give those with timely knowledge enough freedom to act. For example, a minor gasket problem caused the loss of the space shuttle Challenger, resulting in major setbacks for NASA and the United States. This despite the fact that both the problem and the magnitude of the consequences were known. The engineers had warned their management, and had been overruled, dismissed, and disciplined.

Finally, but perhaps most important for a society trying to make a turn around, is the need for economy in action. Action consumes scarce resources. Therefore action must be focused, disciplined, and economically executed. Often the most important choices are what **not** to do.

The nations and the companies that are winning in the markets that matter are being guided by the philosophies of action described above. This is the new model for a successful post-industrial society, and it sets the models for productive, enlightened work forces. This is, we suggest, a more advanced system of management — a better technology — than the West's outdated linear models of regulatory bureaucracies, rules, workers, and bosses.

A Call to Action

Many of the issues we discuss in the chapter could be branded as "political." Perhaps so, but business success and societal prosperity is linked to infrastructure and institutions. If we continue to relegate such issues to our present political system the likelihood is that the status quo will continue. What is needed is a simple list of a few rallying issues that business and the bulk of our citizens, regardless of their party affiliation or political views, can enthusiastically embrace. We suggest the following:

1. A return to basic Constitutional government in the form which is the envy of the world. The current situation is clearly not what the framers of our Constitution envisioned.
2. Campaign reform must be a major priority. Impropriety or the appearance of impropriety in political fund raising should be eliminated from our system. Special interest lobbying is intrinsically antithetical to representative democracy.
3. Education is a major problem area, so educational institutions should be focused exclusively on education. Allowing competition in the educational system seems inevitable and is likely to improve performance. After all, who should be trusted to see children are educated — the government or their parents?
4. There are legitimate "whats" that are owned by the government sector. These issues should be acknowledged. Only government, for example, should put limits on pollution or collect taxes. Where we often tend to get into trouble is when bureaucrats are allowed to dictate the "hows."

    (a) This seems to imply that all government agencies should have oversight boards made up of a representative cross section of all those impacted (citizens, experts, educators, and business people not connected with the government). These oversight boards could report to the only organization permitted to make federal laws, the legislative branch specified in the Constitution.

(b) The IRS and Commerce Departments would be good places to start.

5. Appropriate societal "threat metrics" for the Information Age should be tracked and reported. During the Cold War, we counted missiles and warheads. Artificial metrics like "the Dow" were evolved to report Machine Age industrial economics. We suggest a new scoreboard is needed for Information Age global economic war. Factors considered could include high wage job creation, trade data, patent filings, resource depletion, quality of life, and educational metrics.

Changes are needed and the institutional challenges, in the end, will likely prove to be by far the most difficult to address. A revolution in business and government is required. Still, if we work together to address the difficult challenges of institutions and infrastructure, the 21st century can become a new golden age of prosperity. We must stand up to these challenges, for ourselves, for our children, and for our planet.

Machine Age business has come full cycle from miserly Ebenezer Scrooge to cruel "hired gun" CEOs like "Chainsaw" Dunlap.[26] The old templates lead to scarcity thinking with an emphasis on downsizing, cost cutting, fear, and exploitation. This alarms human rights groups and environmental activists, and justifies ever more powerful regulatory bureaucracies to "protect" workers and society. Executives are painted as villains by the media and in works of fiction. That path leads to decline. Mankind has better options.

In the end, we predict the era of the Machine Age will end, like the Cold War, not with a bang but a whimper. Ironically, the new wave business leaders themselves will replace the old templates with new high prosperity structures where business, society, and employees (no longer rote "workers," but empowered strategic resources) can all win. Shared goals will include high content jobs, increased profits, and prosperity. Firms that cling to the standard routine of the Machine Age will slide to third world wages and thin margins. They won't vanish, but they will become less relevant. This is already happening.

Knowledge-based business taps limitless resource, allowing abundance for those who "own" the knowledge. If the right choices are made in the next few years, a world economy powered by the Engines of Prosperity can allow a new era of civilization. Information Age technology allows a new garden of Eden. Mankind's long dream can at last become reality, as new technology allows a richness of life, wealth creation, freedom, and opportunity beyond our present abilities to even imagine.

Appendix: Monday Morning Actions — Some Concluding Thoughts...

Reflection. The topics we covered in this book are conceptually simple, but hard to do. Most of them deal with values and behavioral change. It is a good time to stand back and reflect on your own core values.

What determines your self-worth? Does being busy and overloaded make you feel important? Is life about how much "headcount" you control? Is a life of control and crisis how you choose to live in five years? In ten? Until you retire? Until the next downsizing? Friends in the outplacement business describe the tragedy of aging managers who devoted their lives to the Machine Age treadmill, only to be cast aside in the end for younger, cheaper, faster treadmill spinners.

Increasingly, outplacement focuses on self-renewal and career transition more than "finding a job." Increasingly, the best and the brightest graduates from the "name" business schools are rejecting traditional careers in the large Machine Age firms for other options.

Learning renewal and behavioral change is now a career and a life skill. Probably nothing is harder for Machine Age managers than learning to delegate and empower. The system of bosses and workers creates a co-dependency of interlocking habits that is hard to break.

Again and again, managers and workers are trained on new and wonderful methods, some of which have merit. In most cases, they go back to the plant and quickly revert to the old ways. When the alarm bells ring, bosses shout orders and workers stop thinking and avoid responsibility.

Leadership and empowerment must be practiced. It is a difficult art, and empowering workers or managers without providing leadership, training, and clear limits is a formula for disaster. Team sports are a better model than Machine Age factories. Teams have to train and practice. They have to win (and lose) against increasingly more difficult opponents in order to learn.

Does your firm have a culture that permits learning mistakes? For example, do you conduct careful reviews of new product development activities and use them for improvement? Is part of management's role developing shared vision and removing barriers? Do you carefully distinguish between learning errors and incompetence? Do you train teams to be winners by letting them fail? Can you accept a temporarily lower level of performance in order to achieve breakthrough results?

For most firms, the answers to all these questions are "no." It is far more common for managers to threaten than lead, and to react rather than think, anticipate, and coach. When the firm gets in trouble, the first thing most do is to downsize. The second action is to assemble the survivors and warn them to work harder, longer, and faster. Aren't these actions what are driving today's business results? *If you keep doing what you are doing, you will keep getting the same results.*

Few CEOs can afford to let all their employees practice for six months, as General Schwartzkopf did. Few dare commit all their resources to any single thing. Still, long journeys start with a single step. Why not take some of your resource, give them shelter, a high value objective, an empowered and trusted leader, and start them training for Information Age success? The trick is to ensure accountability, make the rest of your firm stakeholders in the success, and to combine local empowerment with vision and an overarching strategy. *The leaders are doing these things.*

Endnotes for Chapter 1

[1]"The Blood letting at AT&T is just the Beginning," *Business Week*, January 15, 1996, pg. 30.

[2]John D. Trudel, "The Ruinous Game Called Downsizing," *Upside*, November 1993, pg. 67. Note: Source data comes from broad American Management Association studies that have been repeated and re-validated over many years. Also see, John D. Trudel, "Slashed and Burned Out," *Upside*, July 1994, pg. 10.

[3]Teresa M. Amabile and Regina Conti, "What Downsizing Does to Creativity," Issues and Observations, Volume 15, Number 3, 1995, Center For Creative Leadership, Greensboro, North Carolina. (Note: Dr. Amabile, now at the Graduate School of Business Administration, Harvard University, continues to publish on this subject.)

[4]See for example, Paul Sperry, "Corporate Bulimia," *Investors Business Daily*, April 9, 1996, pg. A4, or Oran Harari, "Ten Reasons why TQM Doesn't Work," *Management Review*, January 1993, pg. 33–38, or "Report Card on TQM," *Management Review*, January 1994, pg. 22–25, or Gilbert Fuchsberg, "Quality Programs Show Shoddy Results," *Wall Street Journal*, May 14, 1992, pg. B1, B9.

[5]From March 3 to 9 of 1996 *NY Times* ran a special series on the societal cost of downsizing. They found that in one third of households a family member had lost their job and 40% more had friends who had lost their jobs. They noted, "More than 43 million (good) jobs have been erased since 1979. Corporate cost-cutting has helped make American business more profitable, but has taken a heavy toll..."

[6]For an intriguing analysis of the U.S. capital structure and its possible impact on competitiveness, see Michael E. Porter, " Capital Disadvantage: America's Failing Capital Investment System," *Harvard Business Review* (September–October 1992): 65–82.

[7]John Trudel, personal communication.

[8]*New York Times* article, September 26, 1995.

[9]John D. Trudel, "The Great Patent Sell-Out," *Upside*, November 1995.

[10]Special Report, "The Case Against Mergers," *Business Week*, October 30, 1995, pp. 122–130.
[11]Frederick Taylor, *The Principles of Scientific Management*. New York: John Wiley, 1917. Taylor believed that management could be made into an exact science. His work focused on routine, repetitive tasks that could be improved with managerial science. Over time, the extension of such thinking moved away from routine behavior to non-managerial tasks, and eventually to activities that were not assigned to executives.
[12]For a good reading on power and politics, see Jeffrey Pfeffer, *Power in Organizations*, New York: Pitman Press, 1986.
[13]John D. Trudel, Private discussions.
[14]Gary Hamel and C.K. Prahalad, Competing for the Future. Boston, Mass.: Harvard Business School Press,. 1994.
[15]Ibid.
[16]Peter M. Senge, *The Fifth Discipline: The Art and Practice of The Learning Organization*, New York: Doubleday, 1990. Also see Senge's follow-on works and John D. Trudel, "Learning Faster than Your Competitors," *Upside*, September 1996, pg. 22.
[17]From Paul Nystrom and William Starbuck, "To avoid organizational crises, unlearn," *Organizational Dynamics* (Spring 1984), pgs. 53–64.
[18]Arie de Geus, *The Living Company*, Boston, Mass.: Harvard Business School Press, 1997.
[19]Kevin J Clancy and Robert S. Schulman, *Marketing Myths that are Killing Business*, McGraw Hill, 1994.
[20]Richard D'Aveni, *Hypercompetition*. New York: Free Press, 1994.
[21]Kenichi Ohmae. *The Borderless World*. New Jersey: Pitman Press, 1992.
[22]Stephen R. Covey, *The 7 Habits of Highly Effective People*, New York: Simon and Schuster, 1989.
[23]Adapted from *Building Your Field of Dreams*, by Mary Manin Morrissey, Bantam Books, 1996.
[24]Nystrom and Starbuck, op cit.
[25]Peter Senge, *op. cit.*
[26]Attribution theory was developed by Harold Kelley. See Harold Kelley and John Michela, "Attribution theory and research." *Annual Review of Psychology*, 1980, 31: 457–501.
[27]"Jack Welch's Encore,"*Business Week* (October 28, 1996), pgs. 155–160.

[28]In the early 1960s, Harvard sociologist Daniel Bell coined the term "post-industrial society" to denote the end of the industrial era and the advent of a service-oriented economy. Such changes were measured by Marc Porat in a classic, *The Information Economy*, 9 vols. Washington, D.C.: Government Printing Office, 1977. Since then, the term has become commonly understood as a service-economy. Our usage of the terms, Machine and Information Age, accentuate the birth of the computer chip and the pervasiveness of its effects. A more contemporary treatise that we have adapted in this section is George Gilder's *Microcosm: The Quantum Revolution in Economics and Technology*, New York: Simon & Schuster, 1989.
[29]This example is borrowed from an editorial by Stanley Schmidt, "Natural Succession," *Analog*, January 1998.
[30]GAO/AIMD-93-2, GAO/AIMD-94-120, GAO/T-AIMD-96-96.
[31]Throughout this book, we will reference a number of books that contrast Newtonian and Quantum physics. For an excellent introduction, see Fritjof Capra, *The Turning Point*. New York: Simon & Schuster, 1982. For a less technical, but equally persuasive book, see Margaret J. Wheatley, *Leadership and the New Science*. San Francisco: Berett-KoehlerPublications, 1994.
[32]Gilder, op cit. pg. 18.
[33]See, for example, W. Brian Arthur, "Increasing Returns and the New World of Business," *Harvard Business Review*, July-August 1996, pp. 100–109.
[34]There are several translations of Sun Tzu's ancient wisdom. Two of the best are: Samuel B. *Griffith, Sun Tzu: The Art of War*, Oxford University Press, 1963. A more contemporary form of the same, derived from Lionel Giles' 1910 English version, is James Clavell, *The Art of War*, Dell Publishing, 1983. The earliest known Western translation (French) was that of Father J. J. M. Amiot, published in Paris in 1772 and again in 1782. It is reported to have been read by Napoleon as a young officer.
[35]John D. Trudel, a private communication from Mr. Robert W. Draeger of Brown and Root Corporation. "Of the top 10 electron tube manufacturers of the 1950s, only three (3) showed up on the 1960 top ten list of semiconductor manufacturers. Of the top 10 integrated circuit manufacturers in 1979, only three (3) appeared on the 1960 list, and only one of these (RCA) was on the 1950s list. Of the top 10 integrated circuit manufacturers in 1986, only four (4) appeared on the 1979 list and *none* were on previous lists."
[36]James C. Collins and Jerry I. Porras, *Built to Last*, New York: Harper Business, 1994.

[37]John D. Trudel, private communication.

[38]Steven Gould and Niles Eldredge, "Punctuated equilibria: The tempo and mode of evolution reconsidered," *Paleobiology*, 3, 1977: 115–151.

[39]Joseph A. Schumpeter, *Capitalism. Socialism, and Economic Performance*. New York: Houghton Mifflin, 1980.

[40]See Paul Fouts, "Acquiring and deploying technology: Strategic perspective on technology evolution," Ph. D. Dissertation. Lundquist College of Business, 1995.

[41]For an excellent comparison between Newtonian and Quantum physics, see Fritjof Capra, *The Turning Point: Science, Society and the Rising Culture*, New York: Bantam Books, 1988.

[42]W. Brian Arthur, "Positive feedbacks in the economy," *Scientific American*, February 1990, pgs. 92–99.

[43]Cited in Alan Meyer, G. Brooks and J. Goes, "Environmental jolts and industry revolutions," *Strategic Management Journal*, 1990, Vol. 11, 93–110; also see Steven Gould and Niles Eldredge, op cit.

[44]Capra op cit.

[45]See Fritjof Capra, *The Web of Life*. New York: Anchor Books, 1996.

[46]Gregory Bateson, *Steps to the Ecology of Mind*. New York: Ballantine, 1972.

[47]Personal communication, Gerardo Ungson.

[48]Arthur op cit.

[49]Michael E. Porter, *Competitive Strategy*. New York: Free Press, 1980.

[50]There have been a spate of books that deal with chaos theory. One by Mitchell is referenced earlier. For an application to business management, see Ralph Stacey, *Managing the Unknowable*. Jossey Bass: San Francisco, 1992.

[51]Paul Romer, "The origins of endogenous growth," *Journal of Economic Perspectives 8*, no. 1, 1994: 2–32.

[52]Pam Woodall, "The World Economy: The Hitchiker's Guide to Cybernomics." *The Economist* (September 28, 1996): Special Guide.

[53]Andy Grove, *Only the Paranoid Survive*, New York: Doubleday, 1996.

[54]*Op cit*. Senge, 1993.

[55]James F. Moore, *The Death of Competition: Leadership and Strategy in the Age of the Business Ecosystem*," New York: Harper Collins, 1996.

[56]Walter Olson, "How Employers are Forced to Hire Murderers and Other Felons, *Wall Street Journal*, June 18, 1997, pg. A23.

[57]Phyllis Schlafly, "The Phillis Schlafly Report," vol. 26, no. 10, May 1993.

Endnotes for Chapter 2

[1]Robert O'Connell, *Of Arms and Men,* New York: Oxford University Press, 1989.
[2]Alvin Toffler, *The Third Wave.* New York: Telecom Library, 1980; James Brian Quinn, *Intelligent Enterprise: A Knowledge and Services Based Paradigm for Industry.* New York: Free Press, 1992; Peter Drucker, *Post-Capitalist Society,* Oxford: Butterworth Heinemann, 1993.
[3]Peter Drucker, *Post-Capitalist Society,* Oxford: Butterworth Heinemann, 1993.
[4]David Ricardo, *The Principles of Political Economy and Taxation.* Homewood, Illinois: Richard D. Irwin, 1967 (first published in 1817).
[5]B. Ohlin, *Interregional and International Trade.* Cambridge, Mass.: Harvard University Press, 1933.
[6]Raymond G. Vernon, Louis T. Wells, Jr., and Subramanian Rangan, *The Manager in the International Economy,* New Jersey: Prentice Hall, 1996.
[7]Gerardo R. Ungson, Richard M. Steers, and Seung Ho Park. *Korean Enterprises: The Quest for Globalization.* Boston, Massachusetts: Harvard Business School Press, 1997.
[8]John D. Trudel, private technology licensing negotiations with a Korean firm.
[9]In fact, in a highly acclaimed book, *The Knowledge Creating Company.* New York: Oxford University Press, 1995, authors Ikujiro Nonaka and Hirotaka Takeuchi argue that creating knowledge will become the key to sustaining a competitive advantage in the future. Another provocative book on this subject of knowledge creation is Taichi Sakaiya, *The Knowledge Value Revolution.* Tokyo: Kodansha International, 1991.
[10]Summarized in Marc Porat, *The Information Economy,* 9 vols. Washington, D.C.: Government Printing Office, 1977.
[11]Ibid.
[12]These examples are drawn from Thomas A. Stewart, *Intellectual Capital: The New Wealth of Nations.* New York: Doubleday Books, 1997, pages 10–11.
[13]Paul Fouts, "The Determinants and Dynamics of the Acquisition and Deployment of Technology," Ph.D. Dissertation, University of Oregon, Lundquist College of Business, Eugene, Oregon1996.
[14]George Stalk, Jr. "Time — The Next Source of Competitive Advantage," *Harvard Business Review,* Reprint 88410.
[15]Pamela Samuelson, "The Copyright Grab," *Wired,* January 1996. Also see,

"Confab Clips Copyright Cartel: How a grab for copyright powers was foiled in Geneva", *Wired*, March 1997.

[16]Charles Platt, "The Great HDTV Swindle," *Wired*, February 1997.

[17]James Brian Quinn, op cit.

[18]Don Tapscott and Art Caston, *Paradigm Shift: The Promise of Information Technology*. New York: McGraw Hill, 1993.

[19]Jim Taylor and Watts Wacher, *Year Delta*, New York: Harper Business, 1997, page 115.

[20]Don Tapscott, *The Digital Economy: Promise and Peril in the Age of Networked Intelligence*. New York: McGraw Hill, 1996.

[21]*Business Week*, October 3, 1994, page 94.

[22]"How H-P Continues to Grow and Grow," *Fortune*, May 2, 1994, pp. 90–100.

[23]Gary Hamel and C.K. Prahalad, *Competing for the Future*, Boston, Mass.: Harvard Business School Press, 1994, Chapter 1.

[24]James Wallace and Jim Erickson, *Hard Drive*, Harper Business, 1992, page 269.

[25]Gary Hamel and C.K. Prahalad, op. cit., page 14.

[26]Sun Tzu, *The Art of War*, Oxford University Press, 1963. (This famous work is the earliest of known treatises on the subject. Various versions are available from several other publishers.)

[27]Peter F. Drucker, "We need to Measure, not Count," *Wall Street Journal*, April 13, 1993.

[28]Peter F. Drucker, op. cit.

[29]Michael E. Porter, "Capital Disadvantage: American's Failing Capital Investment System," *Harvard Business Review*, September–October 1992, pp. 65–82.

[30]Bill Gates, "Not all great software says made in USA," *Oregonian*, February 5, 1995, pg. F5.

[31]Peter Senge, op cit., especially Chapter 14.

[32]John D. Trudel, *High Tech with Low Risk*, 1990, ISBN 0-9626772-2-1, especially page 41. (Now available from The Trudel Group.)

[33]Senge et. al., *The Fifth Discipline Handbook*, 1994, Doubleday, page 45.

[34] Chris Argyris, "Good Communication that Blocks Learning," *Harvard Business Review*, July–August 1994, pp. 77–86.

[35]John D. Trudel, op. cit. page 53.

[36]John D. Trudel, "The Ruinous Game Called Downsizing," *Upside*, November 1993, pg. 67.
[37]Margaret J. Wheatley, *Leadership and the New Science*, Berrett-Kohler Publications, 1992.
[38]Stephen R. Covey, *The 7 Habits of Highly Effective People*, Simon & Schuster, 1989, especially the Chapter on "Principals of Personal Leadership."
[39]Dana Wechsler Linen, "The mother of them all," *Forbes*, January 16, 1995, pp. 75–76.
[40]Fumio Kodama, "Technology Fusion and the New R&D," *Harvard Business Review*, July–August 1992, pp. 70–78. (Kodama's Japanese language book, *Analyzing High Technologies: The Techno-Paradigm Shift*, won the highest awards in Japan, but is unavailable in the West.)
[41]John D. Trudel, op. cit.
[42]John D. Trudel, "The Great Patent Sell-Out, *Upside*, November 1995.
[43]James Wallace and Jim Erickson, op. cit.
[44]James M. Kouzes and Barry Z. Posner, *The Leadership Challenge*, Jossey-Bass, 1987, pp. 57–61.
[45]Regis McKenna, *The Regis Touch*, Addison Wessley, 1985.
[46]John D. Trudel, "Let's Make a Deal," *Upside*, September 1994. See also "Marketing Goes Camping" in the letters section of the November 1994 *Upside*, where Regis McKenna, Inc. explains that "Mr. Trudel Misses a key concept called Camp Marketing" which would "enable the success of the PowerPC architecture…," and Trudel's reply to this letter.
[47]John D. Trudel, op. cit., page 50.
[48]Oren Harari, "The Tarpit of Market Research," *Management Review*, March 1994, pp. 42–44.
[49]John D. Trudel, "When You Can't Manage by the Book," *Upside*, January 1995, page 22.
[50]Margaret J. Wheatley, op. cit.
[51]Gary Zukav, *The Dancing Wu Li Masters*, New York: Bantam Books, 1979.
[52]Geoffrey A. Moore, *Crossing the Chasm*, New York: Harper Business, 1991.

Endnotes for Chapter 3

[1]Michael J. Mandel, "The Real Truth about the Economy: How Government Statistics are Misleading Us," *Business Week*, November 7, 1994, cover story, pp. 110-118.

[2]Leading books on this topic portray the emergence of a global economy linked to global political changes. See Kenichi Ohmae, *The Borderless World: Power and Strategy in the Interlinked Economy*, New York: Harper Business, 1990; Jack Nadel, *Cracking the Global Market, How To Do Business Around the Corner and Around the World*, American Management Association, New York, 1987; Michael E. Porter, *The Competitive Advantage of Nations*. New York: Free Press, 1990; Michael Dertouzos, *What Will Be*. San Francisco: Harper Collins, 1997; and Glen Peters, *Beyond the Next Wave*, London, England: Pitman Publishing, 1996.
[3]This section is based on Paul Dickens, *The Globalization of Industries*. Prentice-Hall: New Jersey, 1993.
[4]Richard D'Aveni, *Hypercompetition*. New York: Free Press, 1995.
[5]See Dickens, op cit.
[6]Ibid.
[7]See Raymond Vernon, Louis Wells, Jr., and Subramanina Rangan, *The Manager in the International Economy*. New Jersey: Prentice Hall, 1996, pgs. 5–9.
[8]For a good discussion, see George Yip, *Global Strategy*. New York: Free Press, 1994
[9]A very good treatment on the subject of globalization and global strategy is George S. Yip, *Total Global Strategy*. New Jersey: Prentice Hall, 1996.
[10]Paul Krugman, *The Geography of Trade*. New York: Free Press, 1992.
[11]Dickens, op cit.
[12]For an excellent update, see Charles Hill, *International Business*. New York: McGraw Hill, 1998.
[13]That the assault of pygmies almost worked is one more example that the rules have changed. Dominant suppliers can be blind-sided by new approaches, both in products and regulatory matters. Processing health care paperwork made Ross Perot's EDS billions, but the money could have gone as easily to IBM.
[14]"America's Choice: High Skills or Low wages?" Report of the Commission on the Skills of the American Workforce, National Center on Education and the Economy, Rochester, New York, June 1990.
[15]Michael Porter, *Competitive Advantage*, New York, Free Press, 1986.
[16]Michael Crichton, *The Rising Sun*. Bantan Press: New Jersey, 1990.
[17]Robert Reich and Eric Mankin, "Joint ventures with Japan give away your future," *Harvard Business Review* (March 1986), 24–44.
[18]See review by Gerardo Ungson, Allan Bird, and Richard Steers, "The

Institutional Foundations of Japanese Industrial and Corporate Policy," Working Paper, Lundquist College of Business, University of Oregon, Eugene, oregon, March 1997.

[19]William Ouchi, *Theory Z: How American Business Can Meet the Japanese Challenge*. Reading, Mass.: Addison Wesley Inc, 1981; Richard Pascale and Anthony Athos, *The Art of Japanese Management*. New York: Simon & Schuster, 1981.

[20]Lester Thurow, *Head to Head*. New York: Morrow Book, 1992.

[21]See Ungson *et al.* Op cit.

[22]Chalmers Johnson, *MITI and the Japanese Miracle*. Stanford, California: Stanford University Press, 1982.

[23]Daniel Okimoto, *Between MITI and the Market*. Stanford, California: Stanford University Press, 1989A.

[24]There are several books that address keiretsu, but the most scholarly to date is Michael Gerlach, *Alliance Capitalism*. Berkeley, California, University of California Press1992.

[25]See Thomas Lifson and Michael Yoshino, *The Invisible Link*. Boston, Mass.: MIT Press, 1984.

[26]Ouchi, op. cit.

[27]This is the subject of many books that include Pascale and Athos, op cit.; Ezra Vogel, *Japan As Number One*, Boston, Mass.: Harvard University Press, 197; James C. Abegglen and George Stalk, Jr. *Kaisha: The Japanese Corporation*. New York: Basic Books, 1985; Rodney Clark, *The Japanese Company*. New Haven: Yale University Press, 1979.

[28]See P.H. Mirvis and E.E. Lawler, "Measuring the financial impact of employee attitude," *Journal of Applied Psychology*, 1977, 62: 1–8.

[29]*Economist*, October 1, 1994.

[30]Kenichi Ohmae, *The End of the Nation State*. New York: Free Press, 1995.

[31]See Peter Mills, *Managing Service Industry: Organizational Practices in a Post-Industrial Economy*. New York: Ballinger, 1986. Mills regularly updates his figures to reflect these changes. See P. Mills, "On the quality of service-encounters," *Journal of Business Research*, 20, 1990: 31–41.

[32]Paul Craig Roberts, "Clinton's Energy Tax: Now That's a Scorched-Earth Policy," *Business Week*, October 27, 1997.

[33]Ibid.

[34]Ibid.

[35]Personal communication, Gerardo Ungson.

[36]This section draws heavily on John Naisbitt, *Megatrends Asia: Eight Asian Megatrends that are reshaping the World*.. New York: Simon & Schuster, 1996; and James Abegglen, *Sea Change: Pacific Asia as the New World Industrial Center*. New York: Free Press, 1994.
[37]Abegglen, ibid.
[38]For an excellent account of Asia's development, see Jim Rohwer, *Asia Rising: Why America Will Prosper as Asia's Economies Boom*. New York: Simon & Shuster, 1996.
[39]Ohmae, op cit.
[40]See Gordon Redding, *The Spirit of Chinese Capitalism*. New York: Walter de Gruyter, 1990; also see Gary Hamilton (ed.) *Asian Business Networks*. New York: Walter de Gruyter, 1996.
[41]Gerardo Ungson, Richard Steers and Seung Ho Park, *Korean Enterprises: The Quest for Globalization*. Boston, Mass.: Harvard Business School Publishing, 1997.
[42]Naisbitt, op cit.
[43]See Gerlach, op cit.
[44]Vogel, op cit.
[45]Ungson et al. op cit.
[46]Porter, *Competitive Advantage of Nations*, op cit.
[47]Personal communication with Dr. Juno Sundaran, Gerardo Ungson.

Endnotes for Chapter 4

[1]"Silicon Valley in Crisis." *Business Week*, March 16, 1985.
[2]For a good discussion from one involved in trade negotiations when the leadership changed, see Clyde Prestowitz, *Trading Places: How we allowed Japan to take the lead*, Basic Books, 1988.
[3]Peter M. Senge, *The Fifth Discipline:* The Art and Practice of The Learning Organization, Doubleday, 1990. Also see Senge's follow-on works and John D. Trudel, "Learning Faster than Your Competitors," *Upside*, September 1996, pg. 22.
[4]See John D. Trudel, *High Tech with Low Risk*, Eastern Oregon State College, 1990, for examples and war stories of cases where good R&D got "stuck" due to corporate politics. (Now available from The Trudel Group.)
[5]John D. Trudel, *High Tech with Low Risk*, op. cit.

[6]This should have been obvious at the time the investments were being made. See, for example, John D. Trudel, "Let's Make a Deal," *Upside*, September 1994, pg. 14. The PowerPC story played out pretty much as this column had predicted.
[7]Norman R. Augustine, "Reshaping an Industry: Lockheed Martin's Survival Story," Harvard Business Review, May-June 1997, pp. 83–94.
[8]Michel Robert, *Product Innovation Strategy Pure and Simple*, McGraw-Hill 1995, pg. 2.
[9]Teresa M. Amabile and Regina Conti, "What Downsizing Does to Creativity," Issues and Observations, Volume 15, Number 3, 1995, Center For Creative Leadership, Greensboro, North Carolina. (Note: Dr. Amabile, now at the Graduate School of Business Administration, Harvard University, continues to publish on this subject.)
[10]Fumio Kodama, op. cit., chapter one.
[11]John A. Byrne, "Re-engineering: What Happened?", *Business Week*, January 30, 1995 (Champy, co-author of with Hammer of the 2 million copy best seller that started the trend admits, "Reengineering is in trouble.")
[12]Barbara Ettorre, "Reengineering tales from the Front," *Management Review*, January 1995.
[13]Michael Hammer, "Who's to blame for all the layoffs?", *Wall Street Journal*, January 22, 1996.
[14]John D. Trudel, the statistical data comes from the Journal of Product Innovation Management, September 1993. A quotation to this effect has been on the back of my business cards for several years.
[15]George Greider, *One World Ready or Not*, Simon & Schuster, 1997. Part Four.
[16]Federal Register, October 1995
[17]Norihiko Shirouzu and Jon Bigness, "7-Eleven Operators Resist System to Monitor Managers," *Wall Street Journal*, June 19, 1997, pp. B1, B3.
[18]From March 3 to 9 of 1996 *NY Times* ran a special series on the societal cost of downsizing. They found that in one third of households a family member had lost their job and 40% more had friends who had lost their jobs. They noted, "More than 43 million (good) jobs have been erased since 1979. Corporate cost-cutting has helped make American business more profitable, but has taken a heavy toll ..."
[19]John A. Byrne, "The Pain of Downsizing," *Business Week*, May 9, 1994, cover story.
[20]John D. Trudel, "The Great Patent Sell-Out," *Upside*, November 1995. See

also, John D. Trudel, "Trudel to Form," Electronic Design columns October 2, 1995, October 24, 1996, November 20, 1995, December 16, 1995, June 9, 1997.
[21]Donald M. Spero, "Patent Protection or Piracy — A CEO views Japan," Harvard Business Review reprint #90511.
[22]John D. Trudel, Electronic Design and Upside columns, op. cit.
[23]Stan Crock and Jonathan Moore, "Corporate Spies Feel a Sting," *Business Week*, July 14, 1997.
[24]John J. Fialka, *War by other means: Economic Espionage in America*, WW Norton and Co., 1997
[25]Greider, *One World, Ready or Not*, op. cit.
[26]Source: American Electronics Association, February 12, 1997, (preliminary data for the year 1995).
[27]For example, the referenced books by Fialka, Prestowitz, and Wolf all make good reading on this topic, op. cit.
[28]Senge, op cit.
[29]Dietrich Dorner, *The Logic of Failure*, Metropolitan Books, New York, 1996.
[30]George Greider, op. cit.
[31]Peter Russell's books focus on this trend. See *The White Hole in Time*, or *The Global Brain*.
[32]John D. Trudel, private discussions with Barbara Marx Hubbard.
[33]Y. Kagita and F. Kodama, NISTEP report No. 15, 1991 (in Japanese).
[34]Kodama, op. cit.
[35]John B. Judis, "How not to conduct trade policy: Flat Panel Flop," *The New Republic*, August 8, 1993.
[36]1997 PDMA International Conference, held in Monterey, "Maximizing the Return on Product Development," PDMA International Office, 401 N. Michigan Avenue, Chicago, IL 60611.
[37]John D. Trudel, *High Tech with Low Risk*, op. cit., pp. 50–65, pp. 105–109.
[38]John D. Trudel, "Trudel to Form," Electronic Design, op. cit.

Endnotes for Chapter 5

[1]See Bruce D. Henderson, "The origin of strategy," *Harvard Business Review* (November–December, 1989), pgs. 139–143.

[2]While there are several treatises relating corporate to military strategy, an excellent summary of the issues is provided by Spyros G. Makridakis, *Forecasting, Planning and Strategy for the 21st. Century*. New York: Free Press, 1990, pgs. 142–162.
[3]B.H. Liddell Hart, *Strategy*. London: Faber & Faber, 1957.
H. Mintzberg and J.B. Quinn, *Strategic Management and Processes*. Englewood, New Jersey: Prentice Hall, 1996.
[4]See Note on Corporate Strategy, Harvard Business School Case. Boston, Mass.: Harvard Business School Clearing House.
[5]For an excellent review of the classical schools of strategic thought, see J.I. Moore, *Writers on Strategy and Strategic Management*, London, England: Penguin Books, 1992.
[6]Much of the discussion here is adopted from Richard Whittington, *What is Strategy and Does It Matter*? London: Routledge, 1993. However, we have taken the liberty of revising his labels for purposes of this book. What Whittington originally terms as "processualists", we have changed to "incrementalism." The reason is that process is also implied in the rational/classical school. In fact, Michel Robert, *Strategy: Pure and Simple*, New York: McGraw Hill, 1993, builds a case for process learning within the rational/classical framework. Moreover, we also changed the original label from "systemic" to "institutional" in that advocates of the other schools of strategy will undoubtedly argue that system-thinking is central to their perspective.
[7]Traditional methods of strategy (i.e., gap models, distinctive competence, etc.) are well illustrated in Igor H. Ansoff's *Corporate Strategy: An Analytical Approach to Business Policy for Growth and Expansion*. New York: McGraw Hill, 1965; and Derek Abell's *Defining the Business: The Starting Point of Strategic Planning*. New Jersey: Prentice Hall, 1980.
[8]For a good discussion, see the following studies on the PIMS (Profit Impact on Market Strategy): Buzzell, R.D. "Product Quality." *PIMS Letter*, No. 4. The Strategic Planning Institute, 1980; Gale, B.T. "Selected findings from the PIMS Project: Market Share Impacts on Profitability," *American Marketing Association Combined Proceedings*, Series 36, 1974, pgs. 471–574; *PIMS, Some Research Findings*, 1977, revised edition. The Strategic Planning Institute, Cambridge, Mass.: 1977. A good summary is provided by P. McNamee, op cit., 1985, pgs. 181–211.
[9]See Gerald Allan, *A Note on the Boston Consulting Group Concept of Comparative Analysis and Corporate Strategy*, ICCH No. 9-175-175, 1976.

[10]The BCG has replaced relative market share with "the size of advantage that can be created over other competitors" and replaced rate of market growth with "the number of unique ways in which that advantage can be created." In doing so, BCG argues that the factors are now expanded and have a wider latitude in their meanings, and also that qualitative skill and judgment are now as important as strict quantitative measures in managing portfolios. See Boston Consulting Group, *Annual Perspective*, 1981.

[11]The concept of experience curve is part of any textbook on corporate strategy. The book, *Strategies for Competitive Success*, by Robert Pitts and Charles Snow. New York: John Wiley & Sons, 1986, has an excellent discussion on pgs. 11–24.

[12]General Electric has developed a nine-cell business portfolio matrix that rates businesses on two factors with three scale-dimensions (strong, average, weak; high, medium, low). The factors are business/strength and competitive position (encompasses the evaluation of relative market share, profit margins, ability to compete on price and quality, knowledge of customer and market, competitive strengths and weaknesses, technological capability, and caliber of management), and long-term product-market attractiveness (encompasses the evaluation of market size and growth rate, industry profit margins, competitive intensity, cyclicality, economies of scale, technology, and social, environmental, legal, and human rights). Based on their respective positions on the matrix, businesses (products) are classified into red, yellow, and green zones, signifying the investment strategies for the firm. See D. Hellreigel and J. Slocum. *Organizational_Behavior*. New York: West Publishing, 1976: 66–67.

[13]Alfred D. Chandler Jr., *Strategy and Structure*. Boston, Mass.: Harvard University Press, 1962.

[14]Larry E. Griener, "Evolution and revolution as organizations grow," *Harvard Business Review*, July–August 1972, 37–46.

[15]Jay Galbraith and Nathanson, *Strategy Implementation: The Role of Structure and Process*. St. Paul, Minn.: West Publishing Company, 1978.

[16]Paul Lawrence and Jay Lorsch, *Organization and Environment*. Boston, Mass.: Harvard University Graduate School of Business Administration, Division of Research, 1967.

[17]M.E. Porter, *Competitive Strategy: Techniques for Analyzing Industries and Competitors* (New York: Free Press, 1980).

[18]Ibid.

[19]Mintzberg suggests some fundamental fallacies of strategic planning: predetermina-tion (that events can be planned in advance); detachment (that strategy is distinct from operations); and formalization (that strategy can be institutionalized). See *Rise and Fall*... pgs. 221–321.
[20]J.B. Quinn, *Strategies for Change: Logical Incrementalism* (Homewood, IL: Irwin, 1980).
[21]Whittington, op cit..
[22]Quinn, op. cit.
[23]Whittington, op cit.
[24]Douglas C. North, *Institutions, institutional change and economic performance*. New York, NY: Cambridge Univ. Press. 1990.
[25]Readings that provide a good understanding of the institutional perspective in the context of Japanese industrial policy include Chalmers Johnson, *MITI and the Japanese Miracle* (Stanford, California: Stanford University Press, 1982); Michael Borrus, James Millstein and John Zysman, *Responses to the Japanese Challenge in High Technology* (Berkeley Roundtable on the International Economy, University of California, Berkeley, 1983). Kenichi Ohmae's representative work is The *Mind of the Strategist* (New York: Ballinger, 1978).
[26]Whittington, op cit.
[27]See Gerardo Ungson, "Perspectives on Interfirm, Global, and Institutional Strategy: International Competition in High Technology," In L. Gomez Meija and M. Lawless (eds.), *Advances in High Technology* (New York: JAI Press, 1992). pgs. 109–125.
[28]Alice Amsden, *Asia's Next Giant: South Korea and Late Industrialization*. New York: Oxford University Press, 1989.
[29]William H. Davidson and Michael S. Malone, *The Virtual Corporation* . New York, HarperCollins, 1992.
[30]Raymond Miles and Charles C. Snow, *Fit, Failure and the Hall of Fame* (The Free Press: New York, 1994); William H. Davidson and Michael S. Malone, *The Virtual Corporation* (New York, HarperCollins, 1992).
[31]For an excellent treatment of resource-based strategy, see David J. Collins and Cynthia A, Montgomery, *Corporate Strategy: Resources and Scope of the Firm*. New York: McGraw Hill, 1997.
[32]C.K. Prahalad and Gary Hamel, "The Core Competence of the Corporation," *Harvard Business Review*, May-June 1990: 79–91.
[33]This is a position taken by Panjat Ghemawat, *Commitment*. New York: Free Press, 1991; and Jay Barney, "Firm resources and sustained competitive advantage," *Journal of Management*, 1991: 99–120.

[34]David Teece, G. Pisano, and Amy Shuen, "Dynamic capabilities and strategic management," *Strategic Management Journal*, 1997.
[35]See C.K. Prahalad, "Strategies for Growth." In Rowan Gibson (ed.), *Rethinking the Future*, London: Nicholas Brealey Publishing, 1997.
[36]Richard D'Aveni, *Hypercompetition*. New York: Free Press, 1995.
[37]Ghemawat, op cit., pg. 122.
[38]This is adopted from Margaret J. Wheatley, *Leadership and the New Science*. San Francisco: Barret-Koehler Publications, Inc., 1994.
[39]This is based in part on H. Richard Priesmeyer, *Organizations and Chaos: Defining the Methods of Nonlinear Management*. Westport, Conn. : Quorum Books, 1992.
[40]C.K. Prahalad, "Strategies for Growth...", op cit., pg. 66.
[41]Gary Hamel, "Reinventing the Basis of Competition." In Rowan Gibson (ed.), *Rethinking the Future*, London: Nicholas Brealey Publishing, 1997.
[42]Priesmeyer, op cit., pg. 197.
[43]Ibid., pg. 198.
[44]Davidson and Malone, op cit.
[45]J.B. Quinn, *The Intelligent Enterprise: A Knowledge and Services Based Paradigm for Industry*. New York: Free Press, 1992.
[46]Charles Handy, *The Age of Unreason*. Boston, Mass.: Harvard Business School Press, 1995.
[47]Charles Handy, *The Age of Unreason*. Boston, Mass.: Harvard Business School Press, 1995.
[48]Raymond E. Miles, Charles C. Snow, John A. Mathews, Grant Miles, and Henry J. Coleman, Jr. "Organizing in the knowledge age: Anticipating the cellular form;" *The Academy of Management Executive*, (November 1997), Volume XI, 4: 7–20.

Endnotes for Chapter 6

[1]The example is taken from "The Technology Paradox," *Business Week*, March 6, 1995.
[2]Michael E. Porter, *Industry Analysis and Competitive Behavior*, Free Press: New York, 1980.
[3]Michael Borrus and John Zysman, "Wintelism and the Changing Terms of Global Competition: Prototype of the Future?" Working Paper 96B,

Berkeley Roundtable for the International Economy. Berkeley; California, February 1997.
[4]Ibid.
[5]Ibid.
[6]Joel Moykr. "The New Information Technology," In Joel Mokyr, *The Lever of Riches: Technological Creativity and Economic Progress.* New York: Oxford University Press, 1990.
[7]"The hitchhiker's guide to cybernomics, *Economist,* September 28, 1996, pg. 8.
[8]For a history, see Boston Consulting Group, *Perspectives on Experience* (Boston Consulting Group, 1968). Also see Boston Consulting Group, *History of the Experience Curve,* Perspective No. 125 (Boston Consulting Group, Boston, Mass., 1973).
[9]Discussed in W. Brian Arthur, op. cit.
[10]A good discussion of Chaos Theory is found in W. Brian Arthur, "Positive Feedbacks in the Economy," *Scientific American,* February 1990: 92–144.
[11]Arthur, B., "Competing technologies, increasing returns, and lock-in by historic events," *Economic Journal,* 1989, 99: 116–131.
[12]Charles W. Hill, "Establishing a standard: Competitive strategy and technological standards in winner-take-all industries," *The Academy of Management Executive,"* Volume XI, 2, May 1997: 7–26.
[13]Arthur, op cit.
[14]P. A. David, Clio and the Economics of QWERTY. *American Economic Review,* 1985, 74: 332–337.
[15]Ibid.
[16]Clayton M. Christensen. *The Innovator's Dilemma: When New Technologies Cause Great Firms to Fail.* Boston, Mass.: Harvard Business School Press, 1997.
[17]Borrus and Zysman, op. cit.
[18]Ibid.
[19]Alfred D. Chandler, *Strategy and Structure.* Cambridge, Mass.: MIT Press, 1962.
[20]Borrus and Zysman, op cit.
[21]Ibid.
[22]Ibid.
[23]Data on this case is lifted from The World VCR Industry, *Harvard Business School Case* 9-387-098, 1987.
[24]There are a number of works that describe semiconductor manufacturing.

A good, concise history is provided in Stan Avgarten. *State of the Art: A Photographic History of the Integrated Circuit*, New Haven, New York: Ticknor and Fields, 1983. A profile of Robert Noyce and Jack Kilby is presented in T. R. Reid, *The Chip*, New York, New York: Simon & Schuster, 1984. The key reference is Rodnay Zaks, *From Chips to Systems*. Berkeley, California: Sybex Incorporated, 1981.

[24]T. R. Reid. *The Chip*. New York, N. Y., Simon & Schuster, 1984, pgs. 66–67.

[25]Ibid.

[26]Ibid., pg. 13.

[27]More detailed history of the evolving industry structure is presented in the following books: *The Semiconductor Industry: Trade Related Issues*, 75775 PARIS CEDEX 16, France: The Organization for Economic Cooperation and Development, 1985; *The Competitive Status of the U.S. Electronics Industry*, Washington, D. C.: National Academy Press, 1984; and Daniel I. Okimoto, Takuo, Sugano, and Franklin B. Weinstein. *Competitive Edge: The Semiconductor Industry in the U.S. and Japan*, Stanford, California: Stanford University Press, 1984.

[28]Reid, T. *The Chip*. New York: Schuster & Schuster, 1984.

[29]See J. G. Abegglen and G. Stalk, *Kaisha: The Japanese Corporation*.
New York: Basic Books, 1985. This is also treated at great length in C. Johnson. "The Institutional Foundations of Industrial Policy." *The California Management Review* (1985), 27, pgs. 59–69.

[30]I. MacKintosh. *Sunrise Europe: The Dynamics of Information Technology*. Oxford, U.K.: Basil Blackwell Ltd., 1986.

[31]Ibid.

[33]Thomas Howell, William Noellert, Janet MacLaughlin, and Alan Wolff, *The Microelectronics Race: The Impact of Government Policy on International Competition*. Boulder, Colorado: The Westview Press, 199, pgs. 4–6.

[34]W. Finan and A. LaMond. "Sustaining U.S. Competitiveness in Microelectronics: The Challenge to U.S. Policy." In Bruce R. Scott and George C. Lodge (eds.) U.S. Competitiveness in the World Economy, Boston, Massachusetts: Harvard Business School Press, 1985, pg. 174.

[35]R. McKenna, S. Cohen and M. Borrus "International Competition in High Technology." *California Management Review*. 1985, 2, pgs. 15–32.

[36]Despite its being widely reported in the press and being totally consistent with administration policy at the time, Boskin now reportedly denies this quote. The remark was apparently made, if it was made, in a private meeting.

Endnotes for Chapter 7

[1]In earlier chapters, we provided key references for books on chaos. For this specific chapter, a reading of the following would be useful: H. Richard Priesmeyer, *Organizations and Chaos: Defining the Methods of Nonlinear Management*. Westport, Connecticut: Quorum Books, 1992; and Mark D. Youngblood, *Life at the Edge of Chaos*. Dallas, Texas: Perceval Publishing, 1997.
[2]See, for example, "GATT: Tales from the Dark Side," *Business Week*, December 19, 1994, pg. 52.
[3]For a good history of McKinsey, see James O'Shea and Charles Madigan, *Dangerous Company: The Consulting Powerhouses and the Businesses They Save and Ruin*. New York: Times Business, 1997.
[4]Also see Michael T. Jacobs, *Short Term America: The Causes and Cures of Our Business Myopia*. Boston, Mass.: Harvard Business School Press. 1991; and Peter Schwartz, *The Art of the Long View*. New York: Doubleday, 1996.
[5]See Tim Jackson, *Inside Intel*. New York: Dutton Book, 1997.
[6]Andy Grove, *Only the Paranoid Survive*. New York: Doubleday, 1996.
[7]For both successful and unsuccessful attempts to do this, see Richard D'Aveni, *Hypercompetition*. New York: Free Press, 1995.
[8]John D.Trudel, *High Tech with Low Risk*, op cit.
[9]George Stalk, Jr. "Time-The Next Source of Competitive Advantage," *Harvard Business Review*, Reprint 88410.
[10]Marketing in cross-linked markets is an emerging topic in corporate and marketing strategy. See M. Porter, *Competitive Strategy*. New York: Free Press, 1980.
[11]For a good account of Microsoft, see Randall E. Stross, *The Microsoft Way*. Reading, Mass.: Addison Wesley, 1997.
[12]See Hazel J. Johnson, *The Banking Keiretsu*. Chicago, Ill.: Probus Publishing Company, 1993; also Kenichi Miyashita and David Russell, *Keiretsu: Inside the Hidden Japanese Conglomerates*. New York: McGraw Hill, 1994.
[13]See Richard Steers, Yoo Keun Shin and Gerardo Ungson, *Chaebol: Korea's New Industrial Might*. New York: HarperBusiness, 1989.
[14]"Jack Welch's Encore," *Business Week* (October 28, 1996), pgs. 155–160.
[15]Personal communication. John Trudel.
[16]Source for Mythical Man Month, XXX
[17]John D. Trudel, *High Tech with Low Risk*, op cit.
[18]John D. Trudel, *High Tech with Low Risk*, op cit.

[19]John D. Trudel, conversations with various Product Development Management Association researchers, most notably Professor Abbie Griffin.
[20]Robert J. Graham and Randall L. Englund, *Creating an Environment for Successful Projects*, Jossey-Bass, 1997.
[21]Stalk, op cit.
[22]Personal communication, John D. Trudel.
[23]See O'Shea and Madigan, op cit.
[24]Dana Milbank, "McKinsey confronts challenge of its own: The burdens of size," *Wall Street Journal*, front page, September 8, 1993.
[25]James O' Shea and Charles Madigan, *Dangerous Company: The Consulting Powerhouses and the Businesses They Save and Ruin*. Time Business, 1997.
[26]John Micklethwait and Adrian Wooldridge, *The Witch Doctors*, Times Books, 1996
[27]Willy Stern, "Did dirty tricks create a best-seller?" *Business Week*, August 7, 1995.
[28]John A. Byrne, "Buzz off my buzzword," *Business Week*, June 5, 1995.

Endnotes for Chapter 8

[1]Jann Mitchell, "Try working overtime on live life instead," *The Oregonian*, March 19, 1995.
[2]Paul Krugman, *The Age of Diminished Expectations*, the MIT press, 1990.
[3]Source: 1992 American Management Association survey. See also John Trudel, "The Ruinous Game Called Downsizing," *Upside*, November 1993, pg. 67.
[4]Peter Russell has written a number of books on this topic. The most recent is *The Global Brain Awakens*, Palo Alto, Global Brain, Inc., 1995.
[5]Peter Russell, *The White Hole in Time*, New York: Harper Collins, 1992.
[6]Gary Putka, "Harvard Dean Defends, Indicts Business Schools," *Wall Street Journal*, March 13, 1995, pp. B1, B8.
[7]Carla Rapoport, "Charles Handy Sees the Future," Fortune, October 31, 1994, pp. 155–168.
[8]James M. Kouzes and Barry Z. Posner, The Leadership Challenge, San Francisco: Jossey Bass, 1987, pp. 47–48.
[9]Charles Handy, *The Age of Paradox*, Boston, Mass., Harvard Business School Press, 1994.
[10]Peter Senge, et. al., *The Fifth Discipline Fieldbook*, Chapter 33, Mental Models.

[11]Mary Mannen Morrissey, lecture at the Living Enrichment Center, Wilsonville, OR, February 17, 1995

[12]For an excellent discussion of the types of business knowledge needed for knowledge-based competition, see Ervin Laszlo and Christopher Laszlo, *The Insight Edge: An Introduction to the Theory and Practice of Evolutionary Management*. Westport, Connecticut: Quorum Books, 1997.

[13]W. Edwards Deming, *Out of the Crisis*, MIT Center for Advanced Engineering Study, Cambridge MA, 1986.

[14]Marvin L. Patterson, *Accelerating Innovation*, Figure 5-2.

[15]Marvin C. Paulk, "U.S. Quality Advances: The SEI's Capability Maturity Model," Software Engineering Institute, Carnegie Mellon University, Pittsburgh, PA, 15213-3890, U.S.A.,

[16]See, for example, John D. Trudel, *High Tech with Low Risk*, pg. 135.

[17]Private discussions between Dick Knight and John D. Trudel

[18]Frank Vaughn, presentation to the Software Association of Oregon Marketing SIG, February 7, 1995.

[19]James A. Belasco and Ralph C. Stayer, *Flight of the Buffalo*, op. cit., pp. 270–272.

[20]Much of the following discussion in this section is based on H. Norman Schwartzkopf with Peter Petre, *It Doesn't Take a Hero*, New York: Bantam Books. 1992.

[21]Ibid.

[22]James M. Kouzes and Barry Z. Posner, op. cit.

[23]See Francis Fukuyama, *Trust: The Social Virtues and the Creation of Prosperity*. New York: Free Press, 1995.

[24]Frederick Betz, *Strategic Technology Management*, New York: McGraw Hill, 1993.

[25]Miyamoto Musashi, *Book of Five Rings*. Woodstock, New York: Overlook Press, 1974.

[26]James M. Kouzes and Barry Z. Posner, op. cit.

[27]James M. Utterback, "Innovation in Industry and the Diffusion of Technology," contained in M. L. Tushman and W. L. Moore, *Readings in the Management of Innovation*. Pitman, 1982, pp. 29–41.

[28]Chee Meng Yap and William E. Souder, "Factor Influencing New Product Success and Failure in Small Entrepreneurial High-Technology Electronics Firms, *J. Product Innovation Management*, November 1994, pp. 418–432.

[29]Lewis M. Branscome, "Does America Need a Technology Policy?" *Harvard Business Review*, March–April 1992, pp. 24–31.
[30]Fumio Kodama, "Technology Fusion and the New R&D," *Harvard Business Review*, July–August 1992, pp. 70–78.
[31]Richard Comerford, "Mecha ... what?," *IEEE Spectrum*, volume 31, number 8, 1994, pp. 46–49.
[32]Fumio Kodama, *Emerging Patterns of Innovation, Harvard Business School Press*, 1995.
[33]Robert D. Hof, "The Education of Andy Grove," *Business Week*, January 16, 1995.
[34]Robert Simons, "Control in an Age of Empowerment," *Harvard Business Review*, March–April 1995, pp. 80–88.
[35]This was a short report on a presentation of Professor Stephen Wheelright, now at Harvard, formerly at Stanford, at the 1994 PDMA International Conference.
[36]John H. Dasburg, "A Taxing Drag on the Airlines," *Wall Street Journal*, March 21, 1995, pg. A14.

Endnotes for Chapter 9

[1]Admittedly, these are stereotyped criticisms of MBA programs that were popular in the mid 1990s. A number of MBA programs have revised their curriculum to meet the requirements of knowledge-based competition. For specifics, see annual reviews of MBA programs published by *Business Week*, *U.S. News and World Today*, and *Forbes* magazines.
[2]For an in-depth treatment of Japanese executives, see Richard Pascale and Anthony Athos, *The Art of Japanese Management*. New York: Warner Books, 1981.
[3]For excellent accounts of Japanese employment practices, see Tadashi Hanami, *Labor Relations in Japan Today*. San Francisco, California: Kodansha, 1981; and Ezra F. Vogel, *Japan as Number*. New York: Harper and Row Publishers, 1979.
[4]Personal communication with Japanese students, Gerardo Ungson.
[5]For a very good book on empowerment and organizations, see Mark D. Youngblood, *Life at the Edge of Chaos*. Dallas, Texas: Perceval Publishing, 1997.

[6]These three steps are adapted John D. Trudel's "Fixing the System" column in *Upside*, August 1995. He got these from a sermon by Mary Manin Morrissey of Living Enrichment Center, Wilsonville Oregon.
[7]For an excellent book on teams, see Lyman D. Ketchum and Eric Trist, *All Teams Are Not Created Equal*. Newberry Park, California: Sage Publishers, 1992.
[8]Stephen P. Covey, *The 7 Habits of Highly Effective People*, op cit.
[9]James A. Belasco and James A. Stayer, *Flight of the Buffalo*, op. cit., pp. 255–256.
[10]A good account of how smaller and more flexible firms are succeeding in the global marketplace, see Charles Garfied, *Second to None*. Homewood, Illinois: Business One Irwin, 1992.
[11]Michael Killen, *IBM: The Making of the Common View*. New York: Harcourt Brace Jovanovich, 1988.
[12]For a good book about Intel, see Tim Jackson, *Inside Intel*. New York: Dutton Publishers, 1997.
[13]See Joseph Blackburn (ed.), *Time-based Competition: The Next Battleground in American Manufacturing*. Homewood, Ill.: Business One Irwin, 1991.
[14]For a good account of McKinsey Consulting, see James O'Shea and Charles Madigan, *Dangerous Company*. New York: Time Business, 1997.
[15]Personal Communication. John D. Trudel.
[16]See Killen, op cit.
[17]Oran Harari and Chip R. Bell, "Is Wile E. Coyote in your office?", *Management Review*, May 1995, pp. 57–62.
[18]For more on this case, see "Trudel's Columns on Management Insights," the first collection, available from The Trudel Group, or "The Ruinous Game called Downsizing," *Upside*, November 1993.
[19]John D. Trudel, private conversations.
[20]Christopher Farrell, et al. "An Old-Fashioned Feeding Frenzy," Business Week, May 1, 1995.
[21]American Management Association survey and Trudel column, op. cit.

Endnotes for Chapter 10

[1]Jean Bodin, *Six Livres de la Republique*, Paris, France, 1576.
[2]This section is adopted from Peter Drucker, *Post-Capitalist Society*. Oxford: Butterworth-Heinneman, 1993.

[3]Ibid.

[4]Joseph Schumpeter, *The Fiscal State*, 1918

[5]Drucker, op cit.

[6]Ibid.

[7]Ibid.

[8]Ibid.

[9]There have been excellent discourses on the relationship between Japanese business and government, but, in particular, see Chalmers Johnson, *The Industrial Policy Debate*. San Francisco: Institute for Contemporary Studies, 1984.

[10]This view of Japanese infrastructure has dominated academic thinking throughout the 19870s to the 1990s. See Johnson op cit.; also Chalmers Johnson, *MITI and the Japanese Miracle*. Stanford, California: Stanford University press, 1982. More recent accounts are more critical of this relationship. For a flavor of such a critique, see Karel van Wolferen, *The Enigma of Japanese Power*. London: MacMillan London Limited, 1989.

[11]Lester Thurow, *Head to Head*. New York: William Morrow & Company, 1992.

[12]This view is attributed primarily to Asian military strategies, such as Sun Tzu. See Chow How Wee, Khai Sheang Lee and Bambang Walujo Hidajat, *Sun Tzu: War and Management*. New York: Addison Wesley, 1996.

[13]See Gerardo Ungson, Allan Bird and Richard Steers, "The Institutional Foundations of Japanese Industrial and Corporate Policy," Working Paper, Lundquist College of Business, University of oregon, Eugene, oregon, March 1997.

[14]There are many books written on Korean culture and management. See Alice Amsden, *Asia's New Giant*. New York: Oxford, 1989; and Gerardo Ungson, Richard Steers and Seung Ho Park, *Korean Enterprises*. Boston, Mass.: Harvard Business School Press, 1997.

[15]Private letter from Senator Thompson to John D. Trudel, August 8, 1997.

[16]John D. Trudel, "The Great Patent Sell-Out," *Upside*, November 1994

[17]Personal communications, John Trudel.

[18]For an excellent review of global labor patterns, see Jeremy Rifkin, *End of Work*. New York: Putnam Publishers, 1995.

[19]Ibid.

[20]See Takeshi Ishida, *Japanese Political Culture*. New Brunswick: Transaction Publishers, 1983; and Naoto Sasaki, *Management and Industrial Structure in Japan*. Oxford: Pergamon Press, 1989.